An introduction to
chemisorption and catalysis
by metals

An introduction to chemisorption and catalysis by metals

R. P. H. GASSER

Formerly University Lecturer in Physical Chemistry and
Fellow of Corpus Christi, Oxford

CLARENDON PRESS · OXFORD · 1985

Oxford University Press, Walton Street, Oxford OX2 6DP

London New York Toronto
Delhi Bombay Calcutta Madras Karachi
Kuala Lumpur Singapore Hong Kong Tokyo
Nairobi Dar es Salaam Cape Town
Melbourne Auckland

and associated companies in
Beirut Berlin Ibadan Mexico City Nicosia

Oxford is a trade mark of Oxford University Press

Published in the United States
by Oxford University Press, New York

British Library Cataloguing in Publication Data
Gasser, R. P. H.
An introduction to chemisorption and catalysis by metals.
1. Heterogeneous catalysis   2. Metal catalysts
3. Chemisorption   4. Surface chemistry
5. Metallic surfaces   6. Gases—
Absorption and adsorption
I. Title
541.3'95   QD505
ISBN 0-19-855163-0

Library of Congress Cataloging in Publication Data
Gasser, R. P. H.
An introduction to chemisorption and catalysis by metals.
Bibliography: p. Includes index.
1. Chemisorption.   2. Catalysis.   3. Metal catalysts.
I. Title.
QD547.G37 1985   541.3'453   84-7954
ISBN 0-19-855163-0

Set and printed in Northern Ireland at
The Universities Press (Belfast) Ltd.

To B

Preface

Like so many scientific books, this one has evolved from a course of lectures. I am indebted to the department of Chemical Engineering at Yale University for the invitation to become Visiting Lecturer and to give a course for graduate students in the early stages of their graduate work.

The level of discussion is intended to allow students with a general knowledge of thermodynamics, kinetics, and quantum theory to reach a position in which they can read and appreciate the substance, though not necessarily the details, of original publications in surface studies. It should, therefore, be possible for senior undergraduates in both Chemistry and Chemical Engineering, as well as graduate students in those subjects, to follow the discussion.

The underlying aim of the book is to rationalize, so far as possible, the wide range of interactions and events which may follow the impact of a gas molecule with a metal surface. The scope, therefore, extends from the physical interactions which lead only to energy exchange, through physical adsorption and chemical adsorption to a discussion of some catalytic processes. However, the subject of heterogeneous catalysis is itself so wide that it has been necessary to be highly selective. The choice of topics for discussion was intended both to illustrate the close conceptual and experimental links between adsorption and catalysis and to complement the contributions from authors dealing with the more immediately practical aspects of catalysis. Two admirable works in this latter field are C. N. Satterfield's *Heterogeneous catalysis in practice* (McGraw Hill, 1980) and B. C. Gates, J. R. Kratzer, and G. C. A. Schuit's *Chemistry of catalytic processes* (McGraw Hill, 1979).

Consonant with the basic idea of the book was the decision to omit any serious discussion of the surface chemical bond. The theory of surface bonding is so difficult that it did not seem possible to bridge the gap between the intended reader and the literature within a reasonable compass. Those who have an advanced knowledge of quantum theory are well catered for in the book edited by T. N. Rhodin and G. Ertl *The nature of the surface chemical bond* (North Holland, 1979).

As far as is practicable, I have tried to show by examples how those actively studying adsorption and catalysis actually go about their experiments and how they interpret the results. However, in an introductory work such as this, it is not appropriate to try and deal with the whole range of techniques currently available. Rather, those which are most frequently encountered in the literature and which have advanced our understanding of the gas–metal interface are discussed.

Any author is indebted to those who have gone before. Whilst a catalogue would be boring it is appropriate to record in particular the pleasure and profit with which I read M. W. Roberts and C. S. McKee's *Chemistry of the metal–gas interface* (Clarendon Press, 1978), F. C. Tompkins's *Chemisorption of gases on metals* (Academic Press, 1978) and the two books on catalytic processes mentioned earlier.

Finally, I am grateful to my colleagues and students, both at Yale and Oxford, with whom I discussed these matters and who very kindly looked at early drafts of the manuscript. Most especially am I indebted to Dr R. M. Lambert of Cambridge University, without whose meticulous reading of the manuscript and encouraging comments the book would not have seen the light of day.

Oxford, Summer 1983   R. P. H. G.

Contents

1. The fundamentals

Introduction

The processes of adsorption and heterogeneous catalysis with which this book is concerned are consequential on the collision of a gas molecule with a solid surface. Following the collision only a limited number of outcomes is possible. These can be listed as follows.

1. The molecule may rebound from the surface; either elastically, i.e. without exchanging energy, or inelastically, i.e. after exchanging energy.

2. The molecule may be adsorbed. Adsorption can be divided into two broad categories:

(a) Physical adsorption, or physisorption, which is associated with the comparatively weak forces of physical attraction such as exist between all molecules (van der Waals forces). Physisorption has a comparatively low enthalpy of adsorption, typically in the region of -40 kJ mol^{-1}.

(b) Chemical adsorption, or chemisorption, which is associated with exchange of electrons and the formation of a chemical bond between the adsorbed gas molecule (the adsorbate) and the metal surface (the adsorbent). As with normal chemical bonds, a wide range of enthalpy changes may occur on chemisorption, though a typical value might be -400 kJ mol^{-1}.

3. Reaction may take place on the surface. The reaction may be solely of the incoming molecule, for example decomposition, or it may be with other species already adsorbed on the surface. This latter possibility may result in disproportionation, of a single adsorbate, or in the production of new species when mixtures of gases are adsorbed.

4. Desorption may occur. This may be of the same species as the incoming molecule or of some new species resulting from chemical reactions at the surface. Desorption may be induced directly by the collision or may be governed by the usual factors which determine the rates of chemical reactions, viz. the concentrations of the reactants and the velocity constant for the reaction. When the chemical reaction takes place more readily on the surface than in the gas phase, that is to say when the rate of formation of product molecules under identical conditions of temperature and pressure is faster, the adsorbent is acting as a catalyst for the particular process. It should be noted that a metal may well act as a catalyst for more than one reaction of the incoming

molecules. In that case a 'selectivity' of the catalyst between two competing reactions may be defined by the ratio of the rates of formation of the two products.

Although a solid surface has some special properties which arise from the discontinuity of the atomic array at the surface, it is useful to begin with some discussion of the ways in which well-established ideas in other areas may help in understanding gas–solid processes. We begin with a consideration of the general features of physisorption and chemisorption.

Intermolecular interactions—physisorption and chemisorption

Physisorption arises from the attractive forces between non-reacting molecules, van der Waals forces. These may be conveniently divided as follows (Atkins 1982).

1. Dispersion forces, also known as London forces. These are universal and arise from instantaneous fluctuations in the electron distribution in an atom or molecule. The resulting temporary electric dipole moment may induce a dipole moment in an adjacent molecule. Interaction between the instantaneous dipoles leads to a force of attraction. The correlated electron fluctuation may be treated by second order perturbation theory and the resulting dispersion energy is of the form:

$$E_1 = -\frac{C}{R^6}$$

where R is the intermolecular separation and the constant C depends upon the ionization potentials and polarizabilities of the molecules concerned.

2. Dipole forces. When a dipolar molecule approaches another molecule, polar or non-polar, there are forces of attraction. If the second molecule is polar the resulting energy of attraction is given by:

$$E_2 \propto -\frac{\mu_1 \mu_2}{R^6}$$

where μ_1 and μ_2 are the dipole moments. If the second molecule is non-polar, an attractive force is produced by the polarization induced in the non-polar molecule (of polarizability α_2) by the polar molecule, and

$$E_3 \propto -\frac{\mu_1^2 \alpha_2}{R^6}.$$

Thus for all these forces, the attractive energy is proportional to R^{-6} and for intermolecular distances of 3 Å (3×10^{-10} m) and typical values of the molecular parameters, the energies of attraction are roughly in the range 1–5 kJ mol^{-1}.

As the molecules approach more closely, forces of repulsion between the electron clouds become dominant. These forces do not lend themselves to quantitative treatment, and their steep rise with decreasing distance is expressed by an empirical function. The best known choice of repulsive function is R^{-12} and is due to Lennard-Jones. The resulting equation for the intermolecular potential energy can be written:

$$E = 4\varepsilon \left[\left(\frac{\sigma}{R} \right)^{12} - \left(\frac{\sigma}{R} \right)^{6} \right]$$

where ε and σ are as defined in Fig. 1.1. This equation is referred to as the Lennard-Jones (12–6) potential. When a molecule approaches a

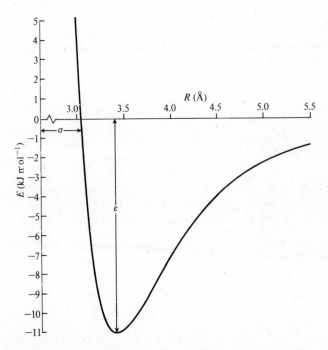

Fig. 1.1. Lennard-Jones (12–6) potential for a system with a well-depth of 11 kJ mol^{-1} and $\sigma = 3.05$ Å. This curve is used for the system (W+O$_2$) shown in Fig. 1.2.

surface the situation is more complex and the potential energy dependence on distance is obtained by summing over all pairwise interactions between the incoming molecule and surface atoms. The result approximates to a (9–3) potential, which is not dissimilar in shape to the (12–6). The above considerations lead to the conclusions that physisorption will have the following general characteristics:

1. it will be non-specific—any gas will adsorb on any solid under suitable circumstances;

2. the heat of adsorption will be of similar order of magnitude to latent heats of evaporation;

3. in view of the previous consideration, extensive physisorption is expected to occur only at low temperatures.

Turning now to chemisorption we notice first that a high degree of specificity is encountered in the reactivity of gas–surface combinations. Thus, not only is there the variation from metal to metal, such as would be anticipated from the differences between the chemistries of the metals, but also different surface planes of the same metal may show considerable differences in reactivity towards a particular gas. One readily discerned thermodynamic reason for such specificity arises when the formation of a surface bond requires the molecule to dissociate. A good example is provided by the dissociative chemisorption of nitrogen on tungsten. The reaction may be written:

$$2W_{surf} + N_{2(g)} = 2(W\!\!-\!\!N)_{surf},$$

and the enthalphy change as:

$$\Delta H_{ad} = D_{N_2} - 2\chi$$

where χ is the strength of the W—N surface bond and D_{N_2} is the dissociation energy of nitrogen. Now the general thermodynamic criterion for a reaction to proceed is that the free energy change must be negative, i.e. in the equation

$$\Delta G = \Delta H - T\,\Delta S$$

ΔG is to be negative. For most types of adsorption, the loss of freedom of movement when a gas phase molecule becomes limited to the two dimensions of a surface leads to a negative entropy change. Thus ΔH_{ad} must normally be negative, i.e. adsorption will usually be exothermic. The possibility of endothermic adsorption can be envisaged in the special case of a molecule which dissociates on adsorption to a highly mobile species on the surface, but that is not the situation for the example we are considering. So for ΔH to be negative:

$$\chi > D_{N_2}/2$$

where D_{N_2}, the dissociation energy of nitrogen, is $941\ kJ\ mol^{-1}$.

The strength of the W—N bond is, therefore, critical in determining whether adsorption will be observed. Even though a particular surface plane (the nomenclature of surfaces will be described in Chapter 4) may form a strong bond to the nitrogen atom, unless this bond energy exceeds (941/2) kJ atom^{-1}, measurable adsorption will not occur. Experimentally it has been found that whereas molecular nitrogen will chemisorb readily on the tungsten (100) plane at room temperature it will not do so on the tungsten (110) plane.

In most gas–metal systems the heat of chemisorption is considerable, even though it may be the balance between two large energies (D and 2χ) in the case of dissociative adsorption, and forms one useful experimental criterion for distinguishing between physisorption and chemisorption. However, there are some borderline cases in which strong physisorption may merge into weak chemisorption and other criteria for distinguishing between them are required.

Once a gas has been adsorbed with a considerable favourable enthalpy change, the van't Hoff equation

$$\frac{d(\ln K)}{dT} = \frac{\Delta H}{RT^2}$$

suggests that high temperatures will be required to remove it. Thus, for example, a temperature in excess of 1000 K is required to desorb all the nitrogen from a polycrystalline tungsten surface.

The general characteristics of chemisorption may be summarized as:
1. a numerically large enthalpy of adsorption;
2. specific to the gas–solid system;
3. layers stable to relatively high temperatures.

Potential energy curves

By analogy with the potential energy curves widely used in molecular spectroscopy to discuss the bonding in diatomic molecules, the chemical bond between an adsorbed species (the adsorbate) and a metal surface (the adsorbent or substrate) could also be represented by a Morse curve. The mathematical expression for the Morse potential is:

$$= D_e\{1 - \exp[-a(R - R_e)]\}^2$$

where D_e is the depth of the potential well binding the species together ($\chi_{W—O}$ in Fig. 1.2), R is the interatomic distance, R_e is the equilibrium separation of the particles, i.e. the bottom of the well, and a is a constant. The general shape of a Morse curve is similar to the Leonard-Jones potential shown in Fig. 1.1. However, two important differences should

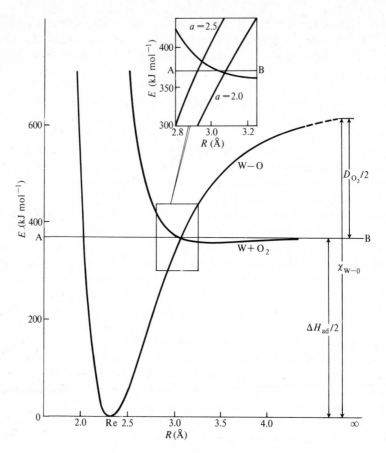

Fig. 1.2. Calculated curves for the physisorption ($W+O_2$) and chemisorption (W—O) of oxygen on tungsten. The insert shows the effect of a change in the shape of the Morse potential, as reflected in the constant a. For $a = 2.0$ the curves cross below the zero-energy line AB, and chemisorption is non-activated. When $a = 2.5$, chemisorption becomes activated.

be noted. First, that the well depth is much greater for the Morse potential, e.g. for HCl, D_e is about 4.6 eV whereas for Ar_2 it is only 0.012 eV. Secondly, the separation at the minimum is much less for the Morse potential. Thus in HCl, R_e is 1.27 Å whereas in Ar_2 it is 3.8 Å.

The implications of these considerations for the interaction between a gas molecule and a solid surface can be expressed on a potential energy diagram, such as Fig. 1.2. On this figure the dissociative adsorption of a molecule is depicted, using for convenience the (12–6) and Morse potentials. For the purpose of exemplification, the dissociative adsorption of oxygen on tungsten has been selected and parameters for the calculation assigned as follows:

1. the constant '*a*' has been given the typical value of two;
2. the equilibrium distance between the oxygen atom and surface tungsten has been taken as 2.32 Å, which is the sum of the oxygen atom single bond radius and half the lattice parameter of tungsten;
3. the depth of the physisorption potential energy well has been taken as 11 kJ mol^{-1}, whilst σ has been assumed to be half the sum of the collision diameter of the oxygen molecule and the lattice parameter of tungsten, i.e. 3.05 Å.

One important feature of potential energy diagrams such as Fig. 1.2 is the point of cross-over from the physisorption to the chemisorption curves. When, as shown for the main curves in Fig. 1.2, transfer from one to the other can occur at an energy below that of the separated molecule and surface, i.e. below AB, no activation energy barrier needs to be overcome for chemisorption to occur. This is indeed true of oxygen chemisorption on tungsten. If, however, the cross-over point is at higher energy, i.e. above AB, as shown in the insert on Fig. 1.2 for $a = 2.5$, chemisorption is activated. Which possibility is realized in practice depends on the shapes and juxtapositions of the physisorption and chemisorption curves, though experience has shown that many of the reactions between clean metal surfaces and simple gases are essentially non-activated. It is of some interest to enquire why this should be so.

For this purpose some analogy may be drawn between a surface reaction and a reaction in the gas phase between an atom or a free radical and a molecule. The characteristic feature of atom reactions is that they tend to have low activation energies. For example, the reaction between hydrogen and bromine in the gas phase takes place by a chain mechanism, the steps being (Gimblett 1970):

$$Br_2 \rightarrow 2Br, \tag{1.1}$$

$$Br + H_2 \rightarrow HBr + H, \tag{1.2}$$

$$H + Br_2 \rightarrow HBr + Br, \tag{1.3}$$

$$H + HBr \rightarrow H_2 + Br, \tag{1.4}$$

$$2Br \rightarrow Br_2. \tag{1.5}$$

The enthalpy changes and activation energies of the individual steps are available. Considering reactions (1.2) and (1.4) together:

$$H + HBr = Br + H_2.$$

The reaction from left to right is strongly exothermic, $\Delta H = -76\,\text{kJ mol}^{-1}$, but has a low activation energy, $5\,\text{kJ mol}^{-1}$. It has therefore, considerable similarities to surface reactions.

If now we imagine a surface being generated by taking a piece of metal and cleaving it down the middle (not often a practicable process for metals, but a common procedure for generating clean oxide or halide surfaces) the metal atoms at the surface will have unused valency electrons. In an idealized experiment, if the metal pieces could be put back together exactly, these electrons would bind them together. Thus a clean metal surface might be regarded as the solid state analogue of an atom or radical and therefore be expected to be highly reactive.

When a solid surface and a gas are in equilibrium at a constant temperature the relationship between the amount of gas adsorbed and the ambient pressure is described as an adsorption isotherm. The measurement of adsorption isotherms is important for the information they yield about the nature of the surface interactions, especially the heats of adsorption. These latter quantities can also be obtained by other methods.

Heats of adsorption and adsorption isotherms

1. Heats of adsorption

The equilibrium between a condensed phase and a vapour can be described by an equation of the form

$$\frac{d(\ln p)}{dT} = \frac{L}{RT^2}$$

where p is the vapour pressure and L is the enthalpy change on evaporation. For a liquid–vapour equilibrium, L is the latent heat of evaporation and the equation is often called the Clausius–Clapeyron equation. When applied to the equilibrium between an adsorbed layer and a gas, the equation must be modified since the enthalpy change is not in general constant, but depends on θ, the fraction of the surface covered. Accordingly, it is necessary to define the coverage when discussing enthalpies of adsorption. The enthalpy change at a particular value of θ is called the isosteric enthalpy of adsorption, ΔH_{st}. On the assumption that ΔH_{st} is independent of temperature over modest temperature ranges, the adsorption equation:

$$\left(\frac{\partial(\ln p)}{\partial T}\right)_\theta = \frac{\Delta H_{st}}{RT^2}$$

can be integrated to give

$$(\ln(p_1/p_2))_\theta = -\frac{\Delta H_{st}}{R}\left(\frac{1}{T_1} - \frac{1}{T_2}\right).$$

If adsorption isotherms are measured at two temperatures a typical result might be as in Fig. 1.3(a). As this figure shows, the pressure required to produce a particular coverage is greater the higher the temperature. Curves such as those in Fig. 1.3(a) allow heats of adsorption to be calculated as a function of coverage; provided always that proper equilibrium is established between the gas and the adsorbed layer. Alternatively, the coverage may be measured at constant pressure whilst varying the temperature. The resulting plot of uptake vs. temperature is called an isobar, and is illustrated in Fig. 1.3(b). Isoteric heats of adsorption are obtained from a series of isobars by reading the values of p and T corresponding to a particular value of uptake, and substituting in the adsorption equation. It is worth noticing that in discussions of gas–solid equilibria, heats of adsorption q_{st} are often used. These are related to ΔH_{st} by

$$q_{st} = -\Delta H_{st}.$$

Heats of adsorption can also be determined by direct calorimetry. The method usually consists of measuring the temperature rise caused by the addition of a known dose of the gas to a film of the metal prepared by evaporation *in vacuo* from a ribbon. A few experiments have been performed using the metal in the form of a fine wire as its own resistance thermometer. The temperature rise caused by admission of a dose of gas is calculated from the resistance change of the wire. Both these experiments yield the differential heat of adsorption, q_d, at the particular value

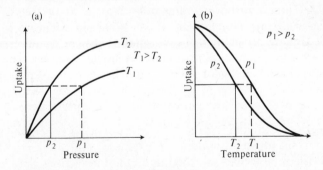

Fig. 1.3. Schematic representation of adsorption. (a) Isotherms. (b) Isobars. ΔH_{st} is calculated from p_1, p_2, T_1, and T_2.

of θ, which is closely similar to the isosteric heat (the difference is RT which is within experimental error).

One further widely used method of obtaining heats of adsorption relies on kinetic rather than thermodynamic data. It is applicable only to those cases of adsorption for which the activation energy is zero. As Fig. 1.2 shows, the heat of adsorption is half the energy difference between AB and the bottom of the potential energy curve (neglecting zero point energy contributions) and this is numerically the same as the activation energy for molecular desorption. Thus if the kinetic rate equation is written in its most general form:

$$\text{rate} = Ac^n \exp(-E_d/RT)$$

where A is the pre-exponential factor, c is the concentration of surface species, n is the order of the surface reaction with respect to the surface species, and E_d is the activation energy for desorption, then:

$$|\Delta H_{ads}| = E_d$$

where $|\Delta H_{ads}|$ is the numerical magnitude of the enthalpy of adsorption. The data necessary for calculating E_d are obtained by rapid heating of the sample, usually in the form of a fine wire or ribbon, to a temperature sufficient to desorb the adsorbed layer rapidly. The pressure burst of desorbed gas is recorded. Analysis of the desorption curve, which will be described in Chapter 3, then yields E_d as well as A and n.

2. Adsorption isotherms

Since adsorption isotherms represent solely the results of experimental observations there is no *a priori* reason why each gas–solid system should not have its own individual shape of isotherm. However, experience has shown that most isotherms fall into one of five main classes (Gregg and Sing 1967). These are shown in Fig. 1.4.

Of these isotherms, classes I and II will be of most concern to us. The first point to note about these two isotherms (which is also true of the other three) is that at low pressures they approach linearity. This may be regarded as a manifestation of Henry's Law, which is of the general form

$$p = \kappa x$$

where κ is a constant and x is the mole fraction in the condensed phase of the component under consideration. At low enough concentrations, where adsorbed molecule–adsorbed molecule interactions are negligible, this equation is obeyed provided that there is no change in molecular complexity (association or dissociation) between the adsorbed state and

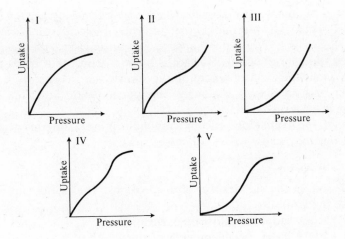

Fig. 1.4. The five adsorption isotherms.

the gaseous state. The second point of interest is that whereas at high pressures curve I approaches a limiting value for the amount adsorbed, curve II indicates an adsorption which can increase indefinitely. These results are interpreted as implying that for curve I the surface becomes saturated with adsorbate and that when this has happened no further uptake can occur. The surface is then said to be covered by a monolayer. The fractional coverage, θ, at any point on the isotherm is defined by:

$$\theta = \frac{\text{Amount adsorbed}}{\text{Monolayer adsorption}}.$$

On the other hand, curve II indicates that many adsorbed layers can be formed. It is, however, possible to make an estimate of the amount of adsorbed gas in the first layer via the BET isotherm discussed later.

Several theories have been advanced to account for the shapes of adsorption isotherms. A feature common to these theories is that assumptions have to be made about the nature of the surface—is it intrinsically homogeneous and if not what is the nature of the heterogeneity?—and about the gas–surface interactions. The strength of these latter may be dependent on the fractional coverage or be independent of it. In practice it is sometimes found that q_{st} declines monotonically with increase in θ. Two main reasons may be advanced why this might happen. When a substrate consists of small particles, for example, it is expected to be intrinsically heterogeneous. Adsorption on the corners and edges is likely to differ from adsorption on the planes. The adsorbing molecules will first be taken up by the most reactive areas, so that the initial value of $|\Delta H|$

will be the maximum. At higher θ the less reactive sites become occupied and $|\Delta H|$ decreases. Secondly, the adsorbed molecules may interact among themselves. When this happens we can envisage the possibility that even though initially the surface is homogeneous, that is to say the enthalpy is independent of where on the surface the gas adsorbs, $|\Delta H_{ads}|$ decreases with increasing coverage. This phenomenon has been called 'induced heterogeneity'.

Some of the commonly encountered adsorption isotherms with the assumptions they make will be discussed next.

Langmuir Isotherm

This is based on the simplest set of assumptions. The model chosen has the following features:

1. the surface is intrinsically homogeneous;

2. the surface has a specific number of sities each of which can adsorb one molecule and when these are all occupied no further adsorption is possible;

3. all the sites are equivalent and the energy of an adsorbed molecule is independent of the presence of other molecules. Condition 2 implies that the uptake is limited to a monolayer, whilst 3 requires that the heat of adsorption is independent of coverage.

It is assumed that when the system has come to equilibrium the rates of adsorption and desorption are equal and not zero; i.e. it is a dynamic equilibrium. In general the rate of adsorption is dependent on:

1. the rate of collision with the surface;

2. a function of the fractional coverage $f(\theta)$;

3. the fraction of molecules possessing the activation energy for adsorption, E_{ad} given by the Boltzmann factor $\exp(-E_{ad}/RT)$;

4. the condensation coefficient, s^*. i.e. the fraction of those molecules with energy $>E_{ad}$ which adsorb.

The kinetic theory of gases gives a value for the rate of collision at pressure p as $p/(2\pi mkT)^{1/2}$, where m is the molecular mass and k is Boltzman's constant. Thus:

$$\text{rate of adsorption} = \frac{p}{(2\pi mkT)^{1/2}} \cdot f(\theta) \cdot s^* \cdot \exp(-E_{ad}/RT).$$

The rate of desorption is dependent on:

1. the specific rate constant for desorption, k_d;

2. a function of the fractional coverage $f'(\theta)$;

3. the fraction of molecules possessing the activation energy for desorption, E_d. So

$$\text{rate of desorption} = k_d f'(\theta)\exp(-E_d/RT).$$

At equilibrium the two rates are equal;

$$p = (2\pi mkT)^{1/2} \frac{k_d}{s^*} \frac{f'(\theta)}{f(\theta)} \exp(E_{ad} - E_d)/RT.$$

In the simplest case of molecular, i.e. non-dissociative, adsorption, $f(\theta)$ is the fraction of vacant sites whilst $f'(\theta)$ is the fraction of occupied sites, i.e.

$$f(\theta) = (1 - \theta), \qquad f'(\theta) = \theta.$$

The energy term in the exponent $(E_{ad} - E_d)$ is the enthalpy of adsorption, ΔH_{ad} (remember that ΔH_{ad} is negative). The isotherm now becomes

$$p = (2\pi mkT)^{1/2} \frac{k_d}{s^*} \cdot \frac{\theta}{1 - \theta} \exp(\Delta H_{ad}/RT) \qquad (1.6)$$

or

$$p = \frac{\theta}{b(1 - \theta)}; \quad \text{where} \quad b^{-1} = (2\pi mkT)^{1/2} \frac{k_d}{s^*} \exp(\Delta H_{ad}/RT).$$

Thus

$$\theta = \frac{bp}{1 + bp}$$

which is the Langmuir isotherm.

This equation can be seen to have the correct limiting behaviour. At low pressures, $bp \ll 1$ and

$$\theta \propto p.$$

At high pressures, $bp \gg 1$ and

$$\theta \rightarrow 1.$$

If two gases A and B are present and competing for the surface, the result will depend upon their pressures and their relative heats of adsorption. If their Langmuir 'b' factors are b_A and b_B, then:

$$\theta_A = \frac{b_A p_A}{1 + b_A p_A + b_B p_B}$$

and

$$\theta_B = \frac{b_B p_B}{1 + b_A p_A + b_B p_B}.$$

When the molecule adsorbs dissociatively into two fragments which occupy single sites at random, the coverage functions are modified as follows:

$$f(\theta) = (1 - \theta)^2 \quad \text{and} \quad f'(\theta) = \theta^2.$$

The justification for these functions is that in adsorption the probability of finding the two empty sites required is the product of the two probabilities of finding one empty site, each of which is $(1-\theta)$. In desorption, two particles must come together to form the desorbing molecule, the encounter probability being proportional to the concentration of each, i.e. to θ^2.

The isotherm for dissociative adsorption then becomes:

$$p = \frac{1}{b}\left(\frac{\theta}{1-\theta}\right)^2$$

or

$$\theta = \frac{(bp)^{1/2}}{1+(bp)^{1/2}}.$$

The low pressure limit, $(bp)^{1/2} \ll 1$, is:

$$\theta \propto p^{1/2}.$$

Other isotherms

One clear weakness of the Langmuir model is the assumption that the heat of adsorption is independent of coverage. In general it is not to be expected that samples such as films prepared by evaporation or the small metal crystallites supported on inert substrates often used for catalytic work will be intrinsically homogeneous. Even in those few cases, such as single crystal surfaces, where the sites may be equivalent initially, lateral interactions between adsorbed species may well alter the heat of adsorption as the coverage increases.

The assumption that adsorption is limited to a monolayer may also be questioned. Its justification will depend in part upon the experimental conditions and in part upon the relative strengths of the interactions between the first layer of adsorbed molecules and (i) the substrate and (ii) the second and subsequent layers of absorbing molecules. In a system for which $|\Delta H_{ads}|$ is considerably larger than the latent heat of evaporation, appreciable adsorption can take place at temperatures well above the boiling point of the adsorbing gas. The tendency to multi-layer formation will be small under these circumstances. On the other hand if a low temperature (i.e. approaching the boiling point) is required to produce significant adsorption into the first layer, this implies that $|\Delta H_{ads}|$ and the latent heat of evaporation are not very different. Temperatures low enough to produce the first layer may also permit other layers to form.

The possibility of a coverage dependence of the heat of adsorption, but not the possibility of multi-layer formation, is taken into account in two other monolayer isotherms. The Freundlich isotherm is of the form

$$\theta = kp^{1/n} \quad (k \text{ and } n \text{ are constants; } n > 1).$$

This equation, originally empirical, can be derived making the assumptions:

1. that the heat of adsorption declines logarithmically with coverage:

$$q = -q_m \ln \theta; \quad (q_m \text{ is a constant})$$

2. that θ has values which neither approach zero or unity; i.e. the isotherm is for intermediate coverages.

The other monolayer isotherm, the Temkin isotherm can be derived if a linearly declining heat of adsorption is assumed, i.e.

$$-\Delta H = -\Delta H_0 (1 - \beta \theta)$$

where ΔH_0 is the initial enthalpy of adsorption. The isotherm is

$$\theta = -\frac{RT}{\beta \Delta H_0} \ln Ap$$

where A is a constant related to the enthalpy of adsorption.

The possibility of multi-layer adsorption is envisaged in the Brunauer–Emmett–Teller (BET) isotherm. The assumption is made that the first layer is adsorbed with a heat of adsorption H_1 whilst the second and subsequent layers are all characterized by heats of adsorption equal to the latent heat of evaporation, H_L. By considering the dynamic equilibrium between each layer and the gas phase the BET isotherm is obtained, viz.

$$\frac{p}{V(p_0 - p)} = \frac{1}{V_m c} + \frac{c-1}{V_m c} \frac{p}{p_0}.$$

In this equation V is the volume of gas adsorbed, p is the pressure of gas, p_0 is the saturated vapour pressure of the liquid at the temperature of the experiment and V_m is the volume equivalent to an adsorbed monolayer. The BET constant c is given by

$$c = \exp(H_1 - H_L)/RT.$$

The BET equation owes its importance to its wide use in measuring surface areas, especially of films and powders. The method followed is to record the uptake of an inert gas (Ar or Kr) or nitrogen at liquid nitrogen temperature ($-195.8\,°C$). A plot of $p/V(p_0 - p)$ vs. p/p_0, usually for p/p_0 up to about 0.3, yields V_m, the monolayer uptake. This value has to be expressed as an area, for which purpose an assumption about the packing of the adsorbed molecules on the surface and the area occupied by each must be made. There is difficulty in obtaining precision in these quantities, but it is often assumed that the molecules are close packed and that a nitrogen molecule occupies $16 \, \text{Å}^2$ or a krypton atom $19.5 \, \text{Å}^2$.

In general the BET isotherm is most useful for describing physisorption, for which H_1 and H_L are of the same order of magnitude; the preceding isotherms are more useful for chemisorption, for which H_1 is usually much greater than H_L. It is worth noting that the BET isotherm reduces to the Langmuir isotherm when $H_1 \gg H_L$.

After this general survey of adsorption we shall return to a more molecular view of the processes associated with the collision of a gas phase molecule with a surface. These processes play a key role in determining the chemical and catalytic activity of the surface, since an inadequate interaction allows the molecule to bounce off the surface. If this happens neither adsorption nor catalytic activity is to be anticipated.

2. Primary processes during collisions

Let us consider first the impact of a gas molecule with a surface free from adsorbed molecules. There is then no complication from the possible effect of a partial adsorbed layer on the behaviour of the remaining uncovered surface. This latter will be the subject of later discussion. The processes following collision will be discussed in turn.

Elastic collisions

For a normal randomly moving gas there is no observed change in gas or surface from elastic collisions. However, if the gas is in the form of a molecular beam and is of not too high a molecular weight, diffraction effects may occur in the reflected beam. These are most readily observed using single-crystal surfaces of alkali halide crystals, where the alternating positive and negative charges provide an effective diffraction grating. Surfaces of metals tend to have a more uniform potential so that diffraction effects from metals are less readily observed.

Inelastic collisions and trapping

The transfer of energy from the gas to the surface, leading to an inelastic collision, is a necessary first step towards adsorption. A modest loss of translational energy by a gas molecule may still leave it with enough energy to escape from the surface, though with a lower velocity. However, if more than some critical amount of energy is lost, the molecule will be unable to climb out of the potential well at the surface and will, therefore, be trapped. Trapping is an essential preliminary to many surface processes and energy transfer is thus a key element in surface reactions. At the stage of being newly trapped the molecule is envisaged as occupying an excited physisorbed state. Having thus become weakly attached to the surface a variety of surface processes is possible. We shall consider these in turn.

Surface processes

The most important processes to follow the trapping of a molecule are the following.

1. Loss of further energy to the surface at the same site.
2. Migration over the surface with loss of energy at other sites.
3. Return to the gas phase by gaining additional energy from the solid.
4. Transfer to the chemisorbed state, either at the initial site (1) or after migration (2).

Once chemisorption has occurred further options exist:

5. Migration of the newly adsorbed molecule or surface reconstruction (i.e. movement of substrate atoms to new positions). These processes are facilitated when the energy liberated by surface bond formation is not immediately dissipated to the lattice.

6. Loss of vibrational energy of the newly chemisorbed species to the lattice with resulting temperature rise.

7. Migration of equilibrated, chemisorbed species.

8. Evaporation either directly from the chemisorbed state or via the

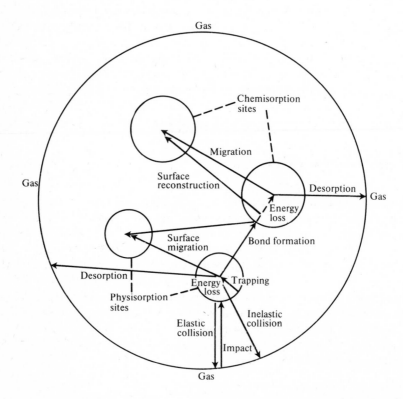

Fig. 2.1. Schematic representation of surface processes. The circles represent the mouths of bell-shaped potential energy wells. The chemisorption well is much deeper than the physisorption well.

physisorbed state. In general, the return to the gas phase is likely to involve the recombination of the atoms of a dissociated diatomic molecule, though an exception is the high temperature (2000 K) desorption of oxygen from tungsten, which occurs as atoms. Polyatomic molecules often decompose during chemisorption and simpler species desorb on heating.

The collision and surface processes discussed thus far can be depicted schematically as shown in Fig. 2.1, where the circles represent potential energy wells which become deeper from circumference to centre. The wells are approximately the shape of inverted church bells, the chemisorption well being much deeper than the physisorption well (as shown in Fig. 1.2).

9. A surface reaction may take place, leading to the evolution of a new molecule. When this reaction occurs more readily on the surface than in the gas phase catalysis is taking place (see Chapter 8).

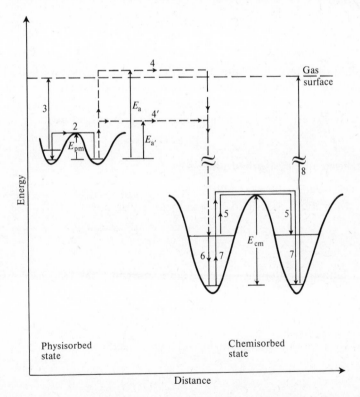

Fig. 2.2. Potential energy profiles for surface processes (not to scale). For explanation, see text. Note that transfer from the physisorbed to the chemisorbed state along path 4 does not require desorption of gas.

The activation energies for the surface processes 2–7 can be characterized by taking a section through the potential energy wells depicted in Fig. 2.1. The result is illustrated in Fig. 2.2. This figure shows that the activation energy for migration in physisorbed or chemisorbed states (E_{pm} or E_{cm}) will be less than the relevant heat of adsorption, although E_{cm} is much greater than E_{pm}. Transfer from the potential energy minimum of the physisorbed state to the chemisorbed state requires an activation energy E_a which can be related to the detailed potential energy curves shown in Fig. 1.2. Depending on where the curves cross, E_a may be large enough for the reaction to be activated, as shown by the broken line 4 in Fig. 2.2, or the reaction may be overall non-activated as shown by the broken line 4' and barrier E_a' which correspond to the situation illustrated in Fig. 1.2.

Each of these processes involves a single adsorbed molecule and may therefore appropriately be described by a first-order rate constant k of the form

$$k = \nu \exp\left(\frac{-E}{RT}\right).$$

In this equation, k is the first-order velocity constant (with units of time^{-1}), ν is a frequency factor appropriate to the particular process and E is one of the activation energies in Fig. 2.2. The application of the equation to a physisorbed molecule will be discussed next.

Behaviour of physisorbed molecules

Let us consider the possibilities open to a physisorbed molecule.
  1. Migration, described by a velocity constant

$$k_{pm} = \nu_{pm} \exp\left(\frac{-E_{pm}}{RT}\right).$$

  2. Transfer to the chemisorbed state, for which the velocity constant is

$$k_a = \nu_a \exp\left(\frac{-E_a}{RT}\right).$$

  3. Desorption, for which the velocity constant k_d is

$$k_d = \nu_d \exp\left(\frac{-E_d}{RT}\right)$$

where $E_d = |\Delta H_{ad}|$, since physisorption is non-activated.

Thus a physisorbed molecule may leave this state either by desorption or by becoming chemisorbed. The effect of the migration option 1 is to increase the chances of the latter process by allowing more than one opportunity for chemisorption during the lifetime of the molecule on the surface. How often the molecule finds itself in a physisorption site adjacent to a chemisorption site before it desorbs will depend principally on the relative magnitudes of the activation energies for migration and desorption.

In order to make an order-of-magnitude estimate of the number of sites a molecule might be expected to visit during its lifetime in the physisorbed state, we note first that the movement of the molecule over the surface is described by the Einstein formula

$$\bar{x}^2 = D\tau$$

where $\bar{x}$ is the mean distance traversed, D is the diffusion coefficient and τ is the lifetime on the surface. The diffusion coefficient can be written in a form similar to the rate constant for migration, namely

$$D = a^2 \nu_{pm} \exp\left(\frac{-E_{pm}}{RT}\right)$$

where a is the distance between sites on the surface. Thus

$$\bar{x}^2 = \tau a^2 \nu_{pm} \exp\left(\frac{-E_{pm}}{RT}\right).$$

The lifetime on the surface can be taken as the characteristic time for the first-order desorption process, which to a first approximation can be taken as k_d^{-1} ($t_{1/2} = \ln 2/k$ for a first-order process). So

$$\bar{x}^2 \approx a^2 \frac{\nu_{pm}}{\nu_d} \exp\left(\frac{|\Delta H_{ad}| - E_{pm}}{RT}\right).$$

The two pre-exponential factors relate to the vibrational motion in the physisorbed state. They are expected to be of a similar order of magnitude, so that

$$\bar{x} \sim a \exp\left(\frac{|\Delta H_{ad}| - E_{pm}}{2RT}\right).$$

For sites of separation a, the average number visited, m is

$$m = \bar{x}/a.$$

It is now possible to make an estimate of the value of m. For the case of oxygen on tungsten we have already used a value of $|\Delta H_{ad}| = 11 \text{ kJ mol}^{-1}$. The activation energy for migration in the physisorbed state (strictly for migration on oxygen-covered tungsten rather than the clean metal) is 3.7 kJ mol^{-1}.

Substituting these values for room temperature gives

$$\ln m \approx \frac{11\,000 - 3700}{2 \times 8.31 \times 298}$$

$$m \approx 4.4.$$

The equilibrium coverage of the physisorbed state is normally very low for gas–metal systems at room temperature and above. The molecule hopping over the surface and visiting the several sites suggested by our rough calculation of m is, therefore, described as being in a 'precursor' state, rather than as being physisorbed. The implications of this precursor state model for the kinetics of adsorption and desorption will be discussed in Chapter 3.

After this phenomenological description of the molecular processes it is time to discuss in more detail the initial process of energy exchange and trapping during a collision.

Gas–surface collisions

Any discussion of the initial exchange of energy between an incoming molecule and a surface is faced with two difficulties. The first is how to measure experimentally the energy transferred, and the second is the choice of a theoretical model for the interaction which is both physically reasonable and mathematically tractable.

On the experimental side there are two principal methods.

1. Accommodation coefficients

To measure accommodation coefficients a fine wire is electrically heated in the gas. Energy is transferred to the gas molecules during collisions with the hot wire. The rebounding gas is thereby heated, but not, in general, to the temperature of the wire. The pressure of gas is chosen so that the mean free path is long compared with the distance from the wire to the vessel wall. There is then no significant loss of energy by the heated molecules to other gas molecules, energy transport being entirely to the walls which are maintained at constant temperature in a thermostat. The accommodation coefficient is defined by

$$\alpha = \frac{T_2' - T_1}{T_2 - T_1} \tag{2.1}$$

where T_1 is the gas temperature, T_2 is the wire temperature and T_2' is the temperature of the rebounding gas, which is such that $T_1 < T_2' < T_2$. The electrical energy required to maintain the wire some $20\,^{\circ}\text{C}$ above the temperature of the gas is measured. Then the rate R of heat transfer, which is also the electrical power supplied to the wire under steady state conditions, is given by

$$R \propto \alpha P(T_2 - T_1).$$

The proportionality constant can be calculated from the kinetic theory of gases. For a gas of molecular weight M, the constant is $7.3 \times 10^3 (MT_1)^{-1/2}$ (in units of $J\,cm^{-2}\,s^{-1}$).

2. Molecular beam scattering

In this experiment a collimated beam of gas molecules produced by effusion through a sequence of orifices hits the surface and the distribution of the scattered molecules is measured with a movable mass spectrometer. More sophisticated experiments use a monoenergetic beam and analyse the energy of the reflected molecules as well as their angular distribution. Three main types of these difficult experiments have so far been performed.

(a) Non-reactive scattering. This is especially associated with inert gases, and nitrogen.

(b) Reactive scattering. The beam consists of either a mixture of reactants, or of one reactant only, the reaction chamber being filled to a convenient (isotropic) pressure of the other reactant. The pressure and temperature dependence of the number of product molecules is measured. In addition, the velocity and angular distribution may be recorded. Examples of systems studied include $H_2 + D_2 \rightarrow HD$ and $CO + O_2 \rightarrow CO_2$.

(c) Diffraction—usually of the light gases helium, hydrogen and deuterium from alkali halide surfaces, although diffraction of helium on a single crystal of tungsten has been observed.

Turning now to the theories of gas–surface energy exchange, we note first that the fundamental process involved is the transfer of the translational energy (in whole or in part) of the gas molecule to the lattice vibrations (phonons) of the solid. The inherent complexity of this system makes rigorous theoretical treatment extremely difficult and some of the simplified models which have been used will be discussed. However, before starting this discussion it will be useful to examine briefly energy exchange between two gas phase molecules, which is much better understood and can give some preliminary insight into the gas–solid interaction.

Energy exchange in the gas phase

The classical theory of translational–vibrational energy exchange is due to Landau and Teller (see Lambert 1977). The model used considers the collision between one rigid molecule CD and one vibrating diatomic molecule AB. The situation is illustrated in Fig. 2.3. As the atom C approaches B along the line of centres (for which the probability of energy exchange is greatest) the interaction between C and B, which gives rise to the energy exchange, depends upon two time-dependent distances. These are the A–B separation, since the molecule AB is vibrating, and the distance between the centres of mass. The intermolecular potential energy, which to a first approximation can be taken as arising from the B–C interaction, is taken to have a distance dependence which is the same as the attractive part of the Lennard-Jones potential discussed earlier (i.e. R^{-6}), but the repulsive R^{-12} term is replaced by an exponential $V(r) = V_0 \exp(-cR)$, where c is a constant.

The position as CD approaches AB is that a non-quantized change in the energy of interaction between C and B is operating on the vibrationally quantized motion of AB at frequency ν. An indication of the probability that this interaction will lead to translational–vibrational energy exchange is given by Ehrenfast's adiabatic principle. This leads to the conclusion that energy exchange will be efficient if the change in force is large during the period of the quantized motion, a result which is expressed via an 'adiabaticity parameter' defined by

$$AP = t_c/t_v$$

where t_c is the 'duration' of the collision and t_v is the period of the vibration. For efficient energy transfer

$$AP < 1; \qquad t_c < t_v.$$

If now the simplifying assumption is made that the short-range repulsive part of the intermolecular potential energy curve dominates the exchange process, the duration of a collision can be taken as the time during which this part operates. If the 'range' of the repulsive forces is a

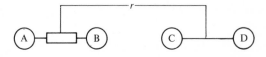

Fig. 2.3. Model for energy exchange: molecule AB is vibrating; molecule CD is rigid.

and the velocity of approach of the molecules is v, then

$$t_c = a/v.$$

The period of oscillation is ν^{-1}, so the requirement for efficient energy exchange becomes

$$a/v < \nu^{-1}; \qquad v > a\nu.$$

Kinetic theory provides the value of v and we substitute this to reach

$$\frac{8kT}{\pi m} > a\nu.$$

Thus efficient exchange is favoured for light molecules with low vibrational frequencies interacting over short distances. The temperature dependence of the probability of exchange predicted by this theory is $\log P \propto T^{1/3}$, and is observed experimentally for many non-polar molecules. A more rigorous quantum-mechanical approach leads to a similar conclusion.

From the point of view of energy exchange at a surface the most important feature of this theory is that the steeper the potential energy curve, that is to say the smaller a, the more efficient the energy exchange becomes. However, an additional feature required for a full theory of gas–solid interactions, which is absent from gas–gas collisions, is the possibility that the participants become attached, i.e. the possibility that trapping can occur must be included. The gas phase analogue of trapping would be the formation of a long-lived gas phase complex between AB and CD. Such complexes are only formed under the special conditions of supersonic beams, when they are known as 'van der Waals' molecules. Examples of van der Waals molecules include Ar_2, $[Ar \cdots HCl]$ and $[Ne \cdots I_2]$.

We shall now consider three approaches to energy exchange and trapping at a surface. Inevitably some limitations are introduced by the simplifications adopted in choosing the models, but this does not prevent useful insights into the factors influencing these processes from being obtained.

Energy exchange at a surface

1. Linear harmonic oscillator chain model

This model (McCarroll and Ehrlich 1963) allows energy transfer and trapping to be discussed separately. The solid is considered to be a chain

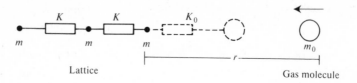

Lattice   Gas molecule

Fig. 2.4. Linear harmonic oscillator model for energy exchange and trapping at a surface. (From McCarroll and Ehrlich 1963.)

of linear harmonic oscillators, with force constant K, to which the incoming particle can become attached, after which it also executes simple harmonic motion with force constant K_0. Initially the chain is not vibrationally excited, i.e. its temperature is 0 K. The lattice particles are of mass m and the gas molecules are of mass m_0. The model is illustrated in Fig. 2.4. To allow escape of the gas molecule the attractive part of the intermolecular potential energy curve terminates at a fixed cut-off value, as illustrated in Fig. 2.5. The results are discussed in terms of two ratios: $\mu = m_0/m$ and $\beta = K_0/K$. The theory is limited to cases for which $\mu < 1$ and $\beta < 1$.

Energy transfer

The repulsive part of the potential, for which $r < r_0$ in Fig. 2.5, is used to discuss energy exchange. The energy exchanged with the lattice at any time t during the collision is given by

$$\left(\frac{\Delta E}{E}\right)_t = 1 - \left(\frac{V_t}{V_0}\right)^2$$

where V_t is the velocity of the particle at time t and V_0 is its initial velocity. The equation of motion of the particle under the influence of the

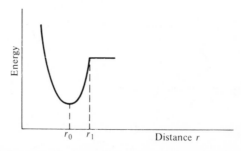

Fig. 2.5. Truncated parabolic potential energy curve for the linear harmonic oscillator model. (From McCarroll and Ehrlich 1963.)

potential gives the velocity ratio as

$$\frac{V_t}{V_0} = 1 - \frac{\beta}{\mu} f(t)$$

in which $f(t)$ is a function of the time elapsed in the collision and is determined by the boundary conditions and includes a power series (Bessel function) in time.

The main point of interest in this result is that the energy exchanged with the surface at any time during a collision depends on the two ratios β and μ and the time, but not on the individual values of the mass or the force constant. The net exchange of energy is then

$$\frac{\Delta E}{E} = 1 - \left(\frac{V_f}{V_0}\right)^2$$

where V_f is the final velocity. The energy ratio $\Delta E / E$ is the accommodation coefficient, as defined by eqn (2.1).

These results show the following.

1. At low values of β the energy exchange increases rapidly as β increases; that is to say, as the steepness of the repulsive potential increases so does the energy exchange.

2. The efficiency increases with the mass of the colliding particle. Thus by the time β and μ have reached unity, which corresponds to a particle colliding with its own lattice, energy exchange is nearly 100% effective.

Trapping

The attractive part of the potential, i.e. between r_0 and r_1 on Fig. 2.5, is now included. The velocity ratio at any time during collision takes the modified form

$$\frac{V_t}{V_0} = 1 - \frac{\beta}{\mu} \{f(t) + Z^{-1} f'(t)\}$$

in which the new term Z^{-1} is given by

$$Z = 2\dot{x}_1 / x_1$$

where $\dot{x}_1$ is the initial relative velocity, $x_1 = r_0 - r_1$, and the additional time-dependent term $f'(t)$ is another Bessel function.

There will be a critical value of the kinetic energy of the incident particle below which trapping occurs. That is to say sufficient translational kinetic energy is transferred to the energy of oscillation to leave the particle with insufficient kinetic energy to return to the cut-off separation r_1, and thus escape. Computation of the critical energy for capture shows

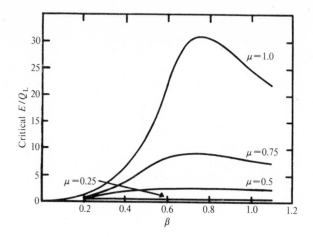

Fig. 2.6. Critical kinetic energy for trapping of a gas molecule on a surface as a function of the force constant ratio $\beta = K_0/K$ for various values of the mass ratio, $\mu = m_0/m$. Q_L is the binding energy of the homogeneous lattice. (From McCarroll and Ehrlich 1963.)

how this energy varies with β and μ. The results, which are shown in Fig. 2.6, are expressed as a function of the ratio of the critical energy E_c to the strength of bonding Q_L of the end lattice particle to its neighbour. There are two main features of this model:

1. the critical energy increases with increase in mass of the particle, i.e. at fixed β, E_c/Q_L increases with μ to a peak at $\mu = 1$ where the ratio exceeds 25;

2. E_c rises with β, so that the deeper the well the greater the critical energy.

This model has the advantage of being conceptually straightforward and gives valuable insights into the factors expected to be important in energy exchange. Its weaknesses arise from the simplifications implied by a one-dimensional lattice which is not vibrating.

2. The hard- and soft-cube models

These models are associated particularly with the interpretation of results from molecular beam experiments. The simpler hard-cube model (Logan and Stickney 1966) provides a useful introduction to the more complex soft-cube model (Logan and Keck 1968).

Hard-cube model

The model is represented in pictorial form in Fig. 2.7. The principal features of this classical model are as follows.

1. There are no forces of attraction between gas and solid.

2. The repulsive interaction between gas and solid is impulsive; that is to say the surface and gas behave as rigid elastic particles.

3. The surface is perfectly smooth and no forces parallel to the surface act on the molecule during the collision.

4. The surface atoms are represented by individual particles confined by square-well potentials (they are in rigid boxes). The incoming molecule interacts by entering the box, colliding with the surface, and leaving.

5. The surface atoms move perpendicular to the surface with a one-dimensional Maxwellian distribution of velocities.

6. The gas molecules also have a Maxwellian velocity distribution.

7. There is a limitation to systems for which $M/m < \frac{1}{3}$.

The aim of the treatment is to determine the angle at which the molecules leave the surface. Any difference between the incident angle θ_i and the reflected angle θ_r must arise solely from a change in the velocity component $u_\perp$ perpendicular to the surface, since for a smooth surface $u_\parallel$ is unchanged on collision. The results are expressed by the angular deviation η of the reflected beam from the specular angle (for which $\theta_r = \theta_i$):

$$\eta = \theta_i - \theta_r.$$

Thus when the gas loses energy to the surface η is negative and vice versa.

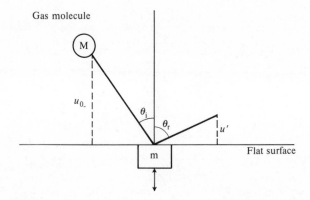

Fig. 2.7. The hard-cube model ($u_{0\parallel} = u'_\parallel$). (From Logan and Stickney 1966.)

Classical mechanics is used to calculate the velocity of the reflected particles in terms of the temperatures T_g and T_s of the gas and the solid respectively and the angle of incidence. It is found that for specular reflection

$$\frac{T_s}{T_g} = \frac{9\pi}{8} \cos^2 \theta_i.$$

When T_s/T_g is less than this value, η is negative for any particular value of θ_i, and vice versa. The basic result of this theory is that if the lattice is 'hot' relative to the gas η will be positive. However, it should be noted that at constant beam temperature an increase in θ_i decreases the relative temperature of the gas since $u_\perp$, which is the important factor in energy exchange, is thereby decreased. A plot of the boundary between regions or positive and negative η is shown in Fig. 2.8. Naturally a distribution of velocities in both beam and surface atom leads to a distribution in the values of θ_r. The shape of the distribution is lobular and as shown in Fig. 2.9(a). θ_r is taken as the angle at which the scattered molecules have their greatest intensity.

The results can be compared with the scattering of argon from silver shown in Fig. 2.9(b). As can be seen the theory is qualitatively satisfactory, but it overestimates the value of η. This result is not unexpected, since in the model an impulsive force between the gas and the surface is used which corresponds to the steepest possible repulsive potential; as we have already noted, the steeper the potential the greater the energy exchange.

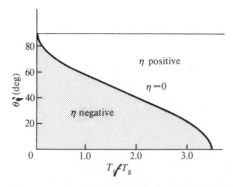

Fig. 2.8. The dependence of the angular deviation η of scattered molecules from the specular angle, on the temperature ratio T_s/T_g and on the angle of incidence θ_i. (From Logan and Stickney 1966.)

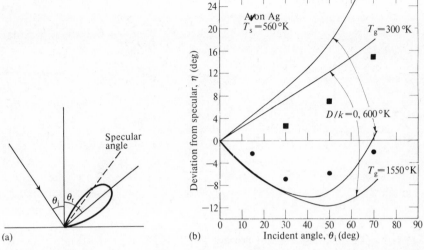

Fig. 2.9. The hard-cube model: (a) the angular distribution of the scattered molecules; (b) comparison of the theoretical predictions for the deviation η of the scattered particles from the specular angle with the experimental results for argon on silver (■, $T_g = 300$ K; ●, $T_g = 1550$ K). (From Logan and Keck 1968.)

Soft-cube model

The soft-cube model is represented schematically in Fig. 2.10. The principal features of the model are as follows:

1. the surface is flat;
2. the gas–surface intermolecular potential is in two parts, an attractive step which accelerates the incoming molecule and an exponential repulsive potential;
3. the surface atom is attached to the lattice by a spring whose surface oscillations are in thermal equilibrium;
4. the gas beam is monoenergetic;
5. there is no limitation on the masses to which it may be applied.

The theory thus involves three variables which can be adjusted to produce the best fit with experiment. These are as follows.

1. The depth D of the attractive step.
2. The characteristic distance b over which the exponential repulsive force operates. The repulsive energy U when the gas molecule is a distance y from the equilibrium position of the surface atom, itself displaced by z, is then given by

$$U = B \exp-\left(\frac{y+z}{b}\right)$$

where B is a constant.

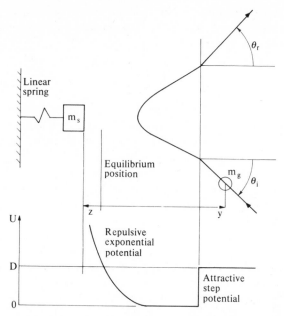

Fig. 2.10. Soft-cube model. (From Logan and Keck 1968.)

3. The oscillation frequency ω of the surface atom.

The calculation of the scattering pattern proceeds by evaluating the force F acting between the gas molecule and the oscillating surface atom as a function of time. From this the impulse imparted to the gas molecule, defined as

$$I = \frac{1}{M_u} \int_{-\infty}^{\infty} F\,\mathrm{d}t$$

where u is the velocity of mass M inside the well and the factor M_u^{-1} makes I dimensionless, is obtained as

$$I = \frac{2\pi}{1+\mu}\frac{\sigma}{\gamma}(1+\tfrac{1}{6}\gamma^2 Z_{OR}) \tag{2.2}$$

where $\mu = M/m$, $\sigma = \omega b/u$, $\gamma = (\pi/2)t_c\omega$ (t_c is the time elapsed in the collision) and Z_{OR} is the displacement of the surface atom at the turning points of the motion. In functional form I is also given by

$$I = 1 + \frac{u'}{u} \tag{2.3}$$

where u' and u are respectively the outgoing and incoming velocities normal to the surface of the gas molecule inside the well. Thus when the molecule is simply turned round without energy exchange $u = u'$ and $I = 2$. In this case $\int_{-\infty}^{\infty} -F \, dt = 2M_u$, the momentum change in reversing direction. However, in general

$$\frac{u'}{u} = \frac{\cot^2 \theta_r / \cot^2 \theta_i + W}{(1 + W)^{1/2}}$$

where $W = 2D/M_{u_\infty}$ and u_∞ is the final velocity of the gas as it leaves the surface. The use of the two equations (2.2) and (2.3) for I to calculate the properties of the reflected beam thus involves the three variables μ, σ, and γ. A comparison of the results for argon scattering from platinum, for xenon on silver, and for argon on silver is shown in Fig. 2.11. As far as the angular distribution is concerned, the general conclusion is that the shape is correctly reproduced, but the theoretical distribution is narrower than the experimental distribution. The reasons for this are probably threefold. First, the experiments include molecules which are trapped and re-emitted. The distribution of these will be random and have a cosine dependence. Secondly, the experiments were performed on Maxwellian and not monoenergetic beams. Thirdly, real surfaces are not entirely smooth.

Trapping can also be calculated from this model using the values of the variables already derived. In general, the fraction trapped is predicted to decrease with increase in T_g and to increase with the mass of the gas molecule.

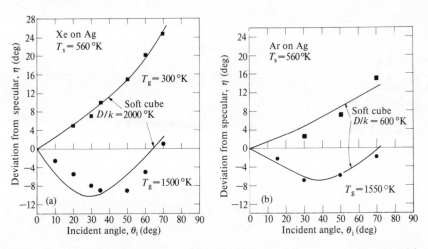

Fig. 2.11. Soft-cube model: comparison of results for (a) xenon on silver and (b) argon on silver, with theory (■, $T_g = 300$ K; ●, $T_g = 1550$ K). (From Logan and Keck 1968.)

3. Square-well model

The aim of this model is to estimate the trapping probability from experimentally measured accommodation coefficients (Weinberg and Merrill 1971). Its features are as follows:

1. an attractive square-well potential of depth D and an impulsive repulsive potential;

2. a smooth surface.

The incoming molecule has kinetic energy ε_i before it reaches the well and therefore hits the surface with total energy $E_i = \varepsilon_i + D$. For trapping, the final energy must be less than the well depth, i.e.

$$\varepsilon_f < D.$$

The energy accommodation coefficient, which is a function of temperature, is

$$\alpha(T) = \frac{E_i - E_f}{E_i - E_s}$$

where E_s is the thermal energy of the solid. Hence, for trapping

$$\frac{D}{E_i} \geq \frac{1 - \alpha(T)}{1 - (E_s/D)\alpha(T)}. \tag{2.4}$$

When the equality in this equation is obeyed, E_i has its maximum value which corresponds to the critical trapping condition. For a defined well depth there is a corresponding critical kinetic energy ε_{ic}, above which the gas molecule is untrapped. Application of the Boltzmann distribution law then gives the trapping probability ξ as

$$\xi = 1 - \exp\left(\frac{-\varepsilon_{ic}}{kT_g}\right)$$

which on substitution from eqn (2.4) and using $E_s = kT_s$ becomes

$$\xi = 1 - \exp\left\{ -\frac{\alpha_c}{1 - \alpha_c} \left(\frac{D}{kT_g} - \frac{T_s}{T_g} \right) \right\}$$

where α_c is the accommodation coefficient at the critical temperature $(D + \varepsilon_{ic})/k$. The maximum value of α can be shown to be

$$\alpha_{max} = \frac{2.4\mu}{(1 + \mu)^2}$$

where

$$\mu = \frac{M_{gas\ atom}}{M_{surface\ atom}}$$

so that the maximum value of the trapping probability is

$$\xi_{max} = 1 - \exp\left\{-\frac{2.4\mu}{1 - 0.4\mu + \mu^2}\left(\frac{D}{kT_g} - \frac{T_s}{T_g}\right)\right\}.$$

For comparison with molecular beam scattering experiments an estimate of D is made from heats of physical adsorption and the observed values of α are used. Satisfactory agreement is achieved for inert gases on tungsten.

4. Scattering and trapping measured by time-of-flight mass spectrometry

The relative importance of (i) elastic scattering, (ii) inelastic scattering, and (iii) trapping followed by desorption have been measured in a series of elegant molecular beam experiments. The systems studied so far have been xenon on platinum (111) and argon and nitrogen on polycrystalline tungsten (Janda, Hurst, Becker, Cowin, Wharton and Auerbach 1980). In this work, attention was focused on the velocity distribution, rather than the angular distribution, of the scattered molecules. The molecular beam was formed by expansion through a supersonic nozzle, a device which generates a beam with a narrow distribution of velocities (equivalent to temperatures between 4 and 73 K) whilst allowing the translational kinetic energy to vary over a wide range (equivalent to temperatures between 300 and 3400 K). Mass spectrometric measurement of the velocity distribution of the scattered molecules, the 'time-of-flight' spectrum, allowed the details of the collision process to be elucidated. Of particular interest is the case of nitrogen on tungsten, where as well as the processes (i), (ii), and (iii) mentioned above, there is also the possibility that trapped molecules may become permanently attached to the surface by chemisorption. We shall consider this example in some detail since the results are germane to both the primary processes discussed in this chapter and the temperature dependence of the rate of chemisorption to be discussed in the next chapter.

The experimental arrangement was that the molecular beam of nitrogen molecules collided at an angle of 45° with a polycrystalline tungsten surface mounted in an ultrahigh vacuum chamber. The time-of-flight spectra of the scattered nitrogen molecules were measured at two angles: (1) 45°, i.e. the specular angle, and (2) 90°, i.e. normal to the surface. The surface temperature T_s ranged between 400 and 1860 K, whilst the beam energy T_B corresponded to temperatures between 290 and 3400 K. The experiments were limited to the early stages of adsorption, a coverage of less than 5 per cent, and were thus characteristic of scattering by a clean tungsten surface.

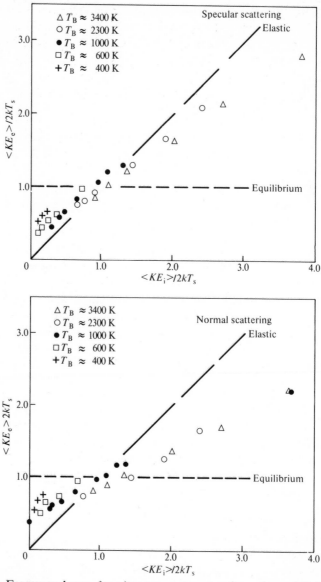

Fig. 2.12. Energy exchange for nitrogen on tungsten from time-of-flight spectra ($\triangle$, $T_B = 3400$ K; $\bigcirc$, $T_B = 2300$ K; $\bullet$, $T_B = 1000$ K; $\square$, $T_B = 600$ K; $+$, $T_B = 400$ K). (a) Scattering at the specular angle. The plot illustrates the linear proportionality between the final kinetic energy $\langle KE_e \rangle / 2kT_s$ and the incident kinetic energy $\langle KE_i \rangle / 2kT_s$. The results expected for no energy exchange are illustrated by the line labelled 'Elastic'; full energy accommodation would give the line labelled 'Equilibrium'. (b) Scattering normal to the surface (otherwise as for (a)). T_B, beam temperature; T_s, surface temperature. (From Kanda *et al.* 1980.)

Considering first energy exchange, two limiting situations can be envisaged. In the first there is no exchange between gas and surface, i.e. elastic scattering. The average energy $\langle KE_e \rangle$ of the exiting scattered molecules is then the same as the average energy $\langle KE_i \rangle$ in the incident beam. As the temperature of the incident beam is raised, the average energy of the scattered beam increases proportionately. At the other extreme the scattered molecules may become completely energy accommodated to the surface. They would then have the average energy appropriate to the velocity distribution at the temperature $2kT_s$ of the surface. The average energy of the scattered beam is then independent of $\langle KE_i \rangle$.

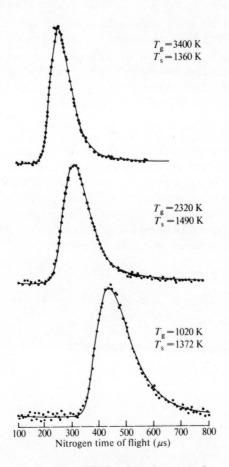

$T_g = 3400 \text{ K}$
$T_s = 1360 \text{ K}$

$T_g = 2320 \text{ K}$
$T_s = 1490 \text{ K}$

$T_g = 1020 \text{ K}$
$T_s = 1372 \text{ K}$

Nitrogen time of flight (μs)

Fig. 2.13. Comparison of time-of-flight data for specular scattering ($\cdot$) with theoretical spectra calculated for direct inelastic scattering (——). (From Janda *et al.* 1980.)

The experimental results, which were expressed in terms of the ratio $\langle KE \rangle / 2kT_s$, lie between the two limits, as shown in Fig. 2.12.

At high beam and surface temperatures (>1000 K) the time-of-flight spectra were consistent with scattering almost exclusively by the direct inelastic scattering process. No evidence either for elastic scattering or for temporary trapping was found. A comparison between spectra predicted for this model and three observed time-of-flight spectra is shown in Fig. 2.13. The curve-fitting procedure involved three separate estimates of the accommodation coefficient. For the two scattered beams, six values of α were thus available; their average was 0.46. Interestingly, this value of α is substantially larger than that for argon scattering by tungsten, which was 0.19.

Turning now to trapping, the results showed that at beam and surface temperatures below 1000 K there was a contribution to the time-of-flight spectra from trapping followed by desorption, as well as from direct inelastic scattering. The measured velocity distribution $I(v)$ was then a linear combination of the contributions $I_I(v)$ and $I_T(v)$ from inelastic and trapping–desorption scattering respectively. Thus

$$I(v) = (1 - F)I_I(v) + FI_T(v)$$

where F is the fraction of the intensity due to trapping–desorption. The calculation of F requires certain reasonable assumptions to be made as to the algebraic form of $I_I(v)$ and $I_T(v)$, and the use of the high temperature values of the accommodation coefficient. This procedure leaves F as the only variable parameter. Some representative values of F are given in Table 2.1 for the specular beam. Anomalous values of F were obtained for the perpendicular beam when the exiting energy of the molecules was comparable with the potential energy well depth. In general, Table 2.1

Table 2.1

Surface temperature (K)	Incident beam energy (K)	F
1240	3360	0.00
1190	1130	0.05
1200	420	0.22
610	2313	0.08
590	1120	0.16
450	3400	0.04
450	1100	0.14
430	290	0.63

illustrates the point that trapping decreases as the temperatures of beam and surface increase. This observation is of fundamental importance to the discussion of the chemisorption of nitrogen on tungsten, since the rate of chemisorption cannot exceed the rate of trapping.

Having thus considered the primary interaction of an incoming molecule with a metal surface, we turn next to the formation of a permanent bond, i.e. to chemisorption.

3. Adsorption and desorption

Adsorption

1. Introduction

As we have already seen, chemisorption of simple gases on clean metal surfaces is frequently non-activated, i.e. E_{ad} is zero. As far as the gas phase is concerned the rate of reaction is therefore determined by the rate of collision with the surface. Under these circumstances the reaction can be looked on as the gas phase analogue of the diffusion-controlled reactions encountered in solution kinetics. However, the spectroscopic and electrochemical methods commonly employed for solutions are not suitable for adsorption reactions. The method most frequently employed for these latter reactions is to reduce the concentration (i.e. pressure) of the gas so much that the rate is slowed down to a level at which it can be followed satisfactorily by mass spectrometry. Typically, a reaction time of not less than a few minutes is desirable and the requirement, as we shall shortly see, is then for a working pressure in the 10^{-8}—10^{-6} Torr range. At these low pressures the reaction rate is first order in gas pressure, but varies with the amount of gas already adsorbed.

The results of adsorption experiments are often expressed by plotting the sticking probability s defined by

$$s = \frac{\text{rate of adsorption of gas}}{\text{rate of collision with surface}}$$

as a function of the fractional coverage θ defined by

$$\theta = \frac{\text{no. of molecules adsorbed}}{\text{maximum uptake}}.$$

It is frequently observed that plots of s versus θ have an initial region in which s is nearly constant, followed by a smooth decline to zero when $\theta = 1$. A typical example of such a sticking probability curve is shown in Fig. 3.1. The value of s as $\theta \rightarrow 0$ is known as the initial sticking probability s_0.

The temperature dependence of s_0 has been investigated for many systems and it is usually found either that s_0 varies little, or that it decreases with increase in temperature.

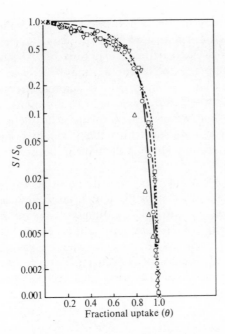

Fig. 3.1. Sticking probability curve for hydrogen on rhodium at 160 K. Experimental results at pressures of 2.2×10^{-8} Torr ($\times$), 2.8×10^{-8} Torr ($\triangle$), 1.7×10^{-8} Torr ($\square$) and 1.6×10^{-8} Torr (∇) are compared with the theory of Kisliuk ($\cdots$) for $s_0 = 0.4$ and $K_1 = -0.42$ and with eqn (3.5) for $K' = 0.1$ and $g(\theta) = (1 - \theta)^2$. (From Edwards *et al.* 1978.)

We have already seen in Chapter 1 why a surface reaction might well be non-activated i.e. why $ds_0/dT \approx 0$. Reactions whose rate decreases with increase in temperature (for a surface reaction this implies that $ds_0/dT < 0$) are rare. One explanation for such a temperature dependence in homogeneous gaseous reactions invokes the postulate of a pre-equilibrium. Thus, for example, in the gas phase reaction between NO and O_2 the velocity constant decreases with increase in temperature. In this case the pre-equilibrium is

$$2NO \rightleftharpoons (NO)_2$$

and is followed by the reaction

$$(NO)_2 + O_2 \rightarrow 2NO_2.$$

It is the movement of the pre-equilibrium to the left as the temperature is increased that reduces the overall rate constant. Similar considerations may apply to chemisorption and to some catalytic processes when the precursor states discussed below are important. Alternatively, the condensation (or trapping) coefficient s^* may decrease with increase in temperature, as we saw in the previous chapter. If condensation is rate determining, such a decrease would lead to a diminution in the sticking probability. Note that we are now drawing a distinction between 'sticking', which implies adsorption into the strongly bonded chemisorbed state, and 'condensation', which is thought of as energy exchange leading to physisorption.

In practice it is not always easy to distinguish between the two possible causes of an s_0 that decreases as the temperature increases. The evidence about the behaviour of nitrogen and tungsten will be discussed later.

The measurement of sticking probability curves is a widely used method of characterizing a reaction between a gas and a metal. The interpretation of the shapes of the curves is therefore of considerable importance and some of the theories advanced to explain these shapes will be outlined next.

2. Precursor theories of chemisorption

There is now general agreement that the characteristic shape of sticking probability curves is to be interpreted as a consequence of the formation of a weakly adsorbed precursor to chemisorption. As we saw in Chapter 2, even the weak forces associated with physisorption may allow a molecule to visit several chemisorption sites during its physisorption lifetime on a surface. Whilst the precursor theory does not equate the attachment of the molecule to the surface in its precursor state with physisorption, the inherent reasonableness of the theory is supported by considerations of physisorption. The precursor model (Kisliuk 1957, 1958) can be represented schematically as

The probabilities of the various steps available to the precursor state molecules are now considered. First it should be noted that these molecules need to be divided into two categories: those which are above a vacant chemisorption site (type 1) and those above a filled chemisorption

site (type 2). The various probabilities can be expressed in tabular form:

Type 1 (above empty site)   Type 2 (above full site)

P_a Probability of chemisorption P'_a
P_d Probability of desorption P'_d
P_m Probability of migration P'_m

Then

$$P_a + P_d + P_m = 1, \; P'_a = 0 \quad \text{and} \quad P'_d + P'_m = 1.$$

Let us consider first a chemisorption process in which the gas molecule occupies a single site on the surface at the stage where the fractional coverage is θ. The probability that a molecule in the precursor state is above an occupied chemisorption site is θ, and the probability that it is above an empty site is $1 - \theta$. The probabilities for the various options, i.e. $(P_a)_1$, $(P_d)_1$ and $(P_m)_1$, are

$(P_a)_1 = P_a(1 - \theta)$ (probability of chemisorption)

$(P_d)_1 = P_d(1 - \theta) + P'_d \theta$ (probability of desorption if above empty site

$+$ probability of desorption if above full site)

$(P_m)_1 = 1 - (P_a)_1 - (P_d)_1$ (probability of migration).

Therefore

$$(P_m)_1 = 1 - P_a - P_d + \theta(P_a + P_d - P'_d).$$

Now consider what may happen at the second site. The molecules that have migrated, with probability $(P_m)_1$, have the same set of possibilities as those available at the first site. The simplifying assumption is made that the second-site probabilities are the same as those of the first site. Then

$(P_a)_2 = (P_m)_1 P_a(1 - \theta)$ (probability of chemisorption)

$(P_d)_2 = (P_m)_1 \{ P_d(1 - \theta) + P'_d \theta \}$ (probability of desorption)

$(P_m)_2 = (P_m)_1 \{ 1 - P_a - P_d + \theta(P_a + P_d - P'_d) \}$ (probability of migration)

$\qquad = (P_m)_1^2$

and so on for third and later sites.

The measured sticking probability $s = \Sigma_i \, P_a$. Therefore

$$s = P_a(1 - \theta)\{ 1 + (P_m)_1 + (P_m)_1^2 + \cdots \}.$$

The series in brackets is $\{1-(P_m)_1\}^{-1}$. Therefore

$$s = \frac{P_a(1-\theta)}{1-(P_m)_1}. \tag{3.1}$$

Now

$$1-(P_m)_1 = P_a + P_d - \theta(P_a + P_d - P_d')$$

and $s = s_0$ when $\theta = 0$. Therefore

$$s_0 = \frac{P_a}{P_a + P_d}. \tag{3.2}$$

Rearrangement of eqn (3.1) yields

$$\frac{s}{s_0} = \left(1 + \frac{\theta s_0}{1-\theta}\frac{P_d'}{P_a}\right)^{-1} = \left(1 + \frac{\theta K}{1-\theta}\right)^{-1} \tag{3.3}$$

where

$$K = \frac{s_0 P_d'}{P_a} = \frac{P_d'}{P_a + P_d}.$$

This theory has just one adjustable parameter, K. As K becomes smaller at a particular value of θ, so the sticking probability ratio s/s_0 increases. Some examples of the dependence of s/s_0 on θ for various values of K are shown in Fig. 3.2(a).

If the molecule requires two sites for adsorption, as for example when a diatomic molecule chemisorbs dissociatively, the equation becomes

$$\frac{s}{s_0} = (1-\theta) \bigg/ \left(1 + \frac{\theta K_1}{1-\theta} + \frac{s_0 \theta^2}{1-\theta}\right); \qquad K_1 = \frac{P_b' - P_a}{P_a - P_b}$$

in which $1 > K_1 > -s_0 > -1$.

There are now two variables K_1 and s_0 and no unique solution is available. Some examples of the plots of s/s_0 versus θ for some representative values of the adjustable parameters are shown in Fig. 3.2(b). The satisfactory fit with the results for the adsorption of hydrogen on polycrystalline rhodium is shown in Fig. 3.1.

Instead of discussing the various surface processes in terms of probabilities it is possible instead to use methods analogous to those of gas phase reactions (Tamm and Schmidt 1970). The new kinetic scheme is

$$A_{(gas)} \overset{s^* f}{\underset{k_d^*}{\rightleftharpoons}} A^* \overset{k_a}{\longrightarrow} A_a$$

where s^* is the sticking probability into the precursor (i.e. the condensation coefficient) and f is the flux of A molecules to the surface. A*

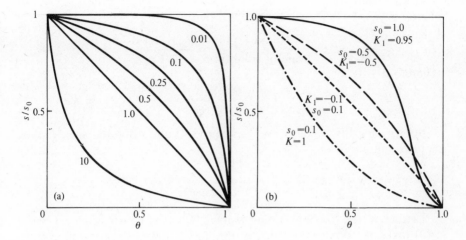

Fig. 3.2. Theoretical sticking probability curves: (a) single-site adsorption according to the equation

$$\frac{s}{s_0} = \left(1 + \frac{\theta K}{1-\theta}\right)^{-1}$$

for various values of K; (b) two-site adsorption according to the equation

$$\frac{s}{s_0} = \frac{(1-\theta)^2}{1-\theta(1-K_1)+\theta^2 s_0}$$

for various values of K_1 and s_0. (From Kisliuk 1957, 1958.)

describes the precursor state and A_a is chemisorbed A. The rate constants for desorption and conversion to the chemisorbed state from the precursor are k_d^* and k_a respectively.

Then for a non-reversible chemisorption step we can write

$$\frac{d[A^*]}{dt} = s^*f - k_d^*[A^*] - k_a[A^*]g(\theta). \tag{3.4}$$

In this equation $g(\theta)$ is the coverage dependence of the probability that an A^* molecule finds a vacant site; for single-site adsorption $g(\theta)$ is taken as $1-\theta$ whilst for two-site adsorption it is $(1-\theta)^2$.

Making the steady state assumption for A^*, since it is in very low concentration, we write

$$\frac{d[A^*]}{dt} = 0.$$

Now the sticking probability as measured experimentally was defined earlier as

$$s = \frac{\text{rate of adsorption}}{\text{rate of collision}} = \frac{k_a[A^*]g(\theta)}{f}.$$

Therefore

$$[A^*] = \frac{sf}{k_a g(\theta)}.$$

Substituting this value of $[A^*]$ in eqn (3.4) and using the steady state approximation yields

$$s = \frac{s^*}{1 + k_d^*/k_a g(\theta)}.$$

Letting $K' = k_d^*/k_a$ and noting that $g(\theta) \to 1$ as $\theta \to 0$ whatever form $g(\theta)$ takes, we obtain the value of the initial sticking probability as

$$s_0 = \frac{s^*}{1 + K'}.$$

Then the equation for the sticking probability curve becomes

$$\frac{s}{s_0} = \frac{1 + K'}{1 + K'/g(\theta)}. \tag{3.5}$$

This equation produced a good fit for the adsorption of hydrogen on the (100) single-crystal plane of tungsten and for hydrogen on polycrystalline rhodium as shown in Fig. 3.1. In both cases the dissociative form

$$g(\theta) = (1 - \theta)^2$$

was used.

3. Temperature dependence of sticking probability curves

There are two aspects of sticking probability curves which may be temperature dependent, the numerical value of the initial sticking probability and the shape of the curve.

A possible interpretation of the temperature dependence in terms of the pre-equilibrium between gas and precursor can be given by combining the probability and rate constant approaches. Thus, if the reasonable assumption is made that the probabilities of the various surface processes are proportional to the appropriate rate constants, we can rewrite eqn (3.2).

$$s_0 = \frac{P_a}{P_a + P_d}.$$

as

$$\frac{k_d}{k_a} = \frac{1-s_0}{s_0}$$

where k_d and k_a are respectively the velocity constants for desorption from the precursor and for chemisorption. Each of these can be expressed by an equation of the form

$$k = \nu \exp\left(\frac{-E}{RT}\right)$$

Then substituting for k_d and k_a and taking logarithms we have

$$\ln\left(\frac{\nu_d}{\nu_a}\right) + \frac{E_a - E_d}{RT} = \ln\frac{1-s_0}{s_0}.$$

Differentiating with respect to $1/RT$ and neglecting any temperature dependence of the pre-exponential factors, the result is

$$\frac{d\{(1-s_0)/s_0\}}{d(1/RT)} = E_a - E_d.$$

The temperature dependence of s_0 thus depends on the difference between the activation energies required for chemisorption (E_a) and desorption (E_d) from the precursor. We have noted that frequently $E_a < E_d$, i.e. overall chemisorption is non-activated. When this is so, the factor $(1-s_0)/s_0$ decreases (i.e. s_0 increases) with increase in $(1/T)$ (i.e. T decreases). That is to say s_0 increases as T decreases. On the other hand, when $E_a > E_d$, s_0 will increase as the temperature increases. In terms of the potential energy diagram (Fig. 1.2) depending on whether the curves intersect above or below the line AB, s_0 is expected to increase or decrease with increase in temperature.

A much studied example of a system in which s_0 decreases with increase in temperature is the adsorption of nitrogen on tungsten. On the assumption that the condensation coefficient s^* did not vary significantly with temperature, a value of $E_d - E_a$ equal to 18 kJ mol^{-1} was derived (King and Wells 1974). However, the validity of the assumption has been called in question by the results of the molecular beam scattering experiments already referred to in Chapter 2 (Janda *et al.* 1980). In this work the early stages of nitrogen adsorption on a polycrystalline tungsten foil were investigated, i.e. the data relate to s_0.

The key observation was the fraction F of the intensity in the scattered beam arising from molecules which had undergone trapping followed by desorption. For a beam of total intensity $I(v)$, the fraction F is related to the direct inelastic contribution $I_I(v)$ (note that there was no direct elastic scattering) and the trapping–desorption contribution $I_T(v)$ by

$$I(v) = (1 - F)I_1(v) + FI_T(v).$$

As the temperature is increased the change in F depends crucially on whether s^* changes. Let us start by considering the implications of the postulated temperature-independent s^*. The intensity of scattered molecules which have undergone trapping–desorption is proportional to $s^* - s_0$ whilst the total intensity of scattered molecules is proportional to $1 - s_0$. Hence

$$F = \frac{s^* - s_0}{1 - s_0}.$$

Before making a comparison between the experimental temperature dependence of F and the value predicted from this equation, there is one complication to be taken into account. The beam results relate to a polycrystalline surface on which the fraction of planes reactive towards nitrogen was not known. However, since s_0 was 0.5, this fraction lay between 0.5 and 1.0. An average value of F was calculated by writing

$$\bar{F} = \frac{s^* - Rs_0}{1 - Rs_0}.$$

where R is the fraction of reactive surface. A plot of $\bar{F}$ for the limiting values of R, 0.5, and 1.0, calculated from the measured sticking probabilities of nitrogen on tungsten (100) is shown in Fig. 3.3. The experimentally observed values of $\bar{F}$ for a beam at 300 K are also shown. As can be seen there is a clear-cut difference between the calculated and

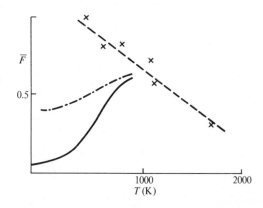

Fig. 3.3. Comparison of the temperature dependence of $\bar{F} = (s^* - Rs_0)/(1 - Rs_0)$ for $R = 0.5$ (- - -) and $R = 1.0$ (——) with the experimental results ($\times$). (From Kanda *et al.* 1980.)

observed trends of $\bar{F}$ with temperature. It thus appears that the decline in s_0 at higher temperatures is not solely due to the movement of the pre-equilibrium between gas phase and precursor state in favour of the gas phase; there is also a contribution from a reduction in the condensation coefficient.

Turning now to the shape of the sticking probability curve, we can use eqn (3.5) to investigate the way in which the shape will change with increase in temperature. In eqn (3.5)

$$\frac{s}{s_0} = \frac{1 + K'}{1 + K'/g(\theta)}$$

we note that K' is the ratio of the velocity constants for desorption and chemisorption from the precursor. Expressing these in the exponential form used above, taking logarithms, and differentiating we have

$$\frac{d(\ln K')}{d(1/RT)} = E_a - E_d.$$

For non-activated adsorption, $E_a < E_d$ and K' increases with increase in temperature. For any particular value of $g(\theta)$ this result implies that s/s_0 decreases as the temperature increases (note that $g(\theta) < 1$). The sticking probability curve thus becomes less convex to the axes as the temperature increases, essentially because the weakly held precursor increasingly tends to desorb rather than become chemisorbed. This is an intuitively reasonable conclusion.

The above considerations thus give a satisfactory account of the frequent observations that s_0 does not increase with increase in temperature and that the influence of the precursor in sustaining the value of s declines as the temperature increases.

Although many systems conform to the general principles discussed above, exceptions have been observed. Some of these arise because there may be coverage-dependent interactions which have not been included in the models used so far. Two such special cases will be discussed.

4. Special cases of sticking probability curves

Nitrogen on tungsten (100)

The need for modification of the model arose because the sticking probability curves were markedly more sigmoid than usual. To account for these shapes the model was modified in two ways.

1. A repulsive pairwise interaction between adjacent chemisorbed nitrogen atoms (the adsorption is dissociative) of magnitude ω was included.

2. The form of $g(\theta)$ was modified using a detailed statistical argument to give

$$g(\theta) = 1 - \theta - \frac{2\theta(1-\theta)}{\{1 - 4\theta(1-\theta)B\}^{1/2} + 1}$$

where $B = 1 - \exp(-\omega/kT)$ and $\theta = N_a/N$. N_a is the number of occupied sites and N is the total number of surface sites.

The resulting expression for the sticking probability curve is

$$\frac{s}{s_0} = \left[1 + K_1\left\{\frac{1}{g(\theta)} - 1\right\}\right]^{-1}$$

where

$$K_1 = \frac{P_d'}{P_a + P_d}.$$

There are now three adjustable parameters N, ω, and K_1. The excellent fit shown in Fig. 3.4 was obtained for $N = 9.8 \times 10^{14}\ \mathrm{cm}^{-2}$ and $B = 0.98$, both of which are independent of temperature, and a value of K_1 increasing from 0.082 at 300 K to 0.5 at 773 K. A further feature of the treatment was the prediction that s_0, which declined from 0.58 at 300 K to 0.22 at 773 K, would become independent of temperature at low temperatures. This has indeed been observed between 77 and 195 K.

Nitrogen on tungsten (110)

Prior to the sticking probability experiments to be described, the main results from studies of the interaction between nitrogen and the single-crystal tungsten (110) plane can be summarized as follows.

1. At room temperature the tungsten (110) plane is much less reactive

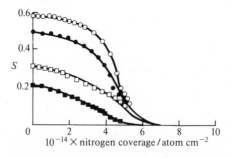

Fig. 3.4. Sticking probability data for nitrogen on tungsten (100):———, theoretical curves; ○, 300 K; ●, 443 K; □, 663 K; ■, 773 K. (From King and Wells 1972.)

than the tungsten (100) plane discussed above. There is even doubt as to whether any adsorption at all occurs. If it does, the initial sticking probability is at most 0.05 and the maximum uptake is about 1×10^{13} molecules cm^{-2}.

2. At low temperatures (90–120 K) the plane becomes more reactive and a surface species designated γ is formed. The main properties of the γ state are that it is weakly held (desorption temperature 190 K), it is not dissociatively adsorbed ($^{14}N_2$ and $^{15}N_2$ when co-adsorbed do not give isotopic mixing), and it sits upright on the surface (by surface photoelectron spectroscopy). The question arose as to whether nitrogen in the γ state on tungsten (110) could convert to a strongly bonded dissociated state, such a transfer having been observed on polycrystalline samples. Although it appeared that transfer did not occur on tungsten (110), the curious behaviour of the sticking probability in the γ state called for a modification of theory. The experimental result at 85 K is shown in Fig. 3.5. After careful analysis of possible experimental artefacts, Bowker and King (1979) concluded that the unexpected maximum in the sticking probability curve was indeed genuine. To account for this result a model with the following characteristics was chosen.

1. The rising sticking probability is attributed to improved transfer from the precursor state to the γ state when a nearest-neighbour γ site is already occupied.

2. The improved transfer arises from an attractive lateral interaction which produces a linear decrease in the energy barrier to transfer from

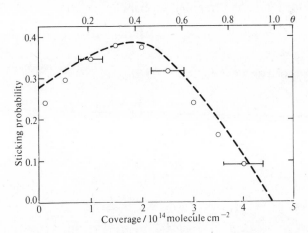

Fig. 3.5. Sticking probability curve for nitrogen on tungsten (110) at 85 K (O) compared with the lateral interaction model (- - -). (From Bowker and King 1979.)

the precursor to the γ state, i.e.

$$E_a = E_a^0 - b\theta$$

3. The trapping probability into the precursor state is 0.9 and is independent of temperature and occupancy of the surface.

The kinetic scheme is then

neighbouring site.

$$\text{gas} \underset{P_d}{\overset{\alpha}{\rightleftharpoons}} \text{trapped precursor} \overset{P_m}{\underset{P_a}{\diagdown}}$$

γ site

The various probabilities are as before, and α is the trapping probability. If we again make the assumption that the probabilities are proportional to the velocity constants

$$\frac{P_d}{P_a} = \frac{\nu_d}{\nu_a} \exp\left(-\frac{E_d}{RT} + \frac{E_a^0 - b\theta}{RT}\right)$$

$$= \frac{\nu_d}{\nu_a} \exp\left(-\frac{\Delta E^0 + b\theta}{RT}\right) \tag{3.6}$$

where $\Delta E^0 = E_d - E_a^0$. Next eqn (3.3) for a non-dissociative sticking probability curve is used in the form

$$s = \alpha\left(\frac{P_a}{P_a + P_d}\right)\left(1 + \frac{\theta K}{1 - \theta}\right)^{-1}. \tag{3.7}$$

The introduction of the trapping probability α gives

$$s_0 = \frac{\alpha P_a}{P_a + P_a}.$$

Substitution of eqn (3.6) in eqn (3.7) yields

$$s = \frac{\alpha}{1 + De^{-c\theta}}\left(1 + \frac{K\theta}{1 - \theta}\right)$$

where

$$D = \frac{\nu_d}{\nu_a} \exp\left(-\frac{\Delta E^0}{RT}\right)$$

$$c = \frac{b}{RT}.$$

A satisfactory fit is produced for $D = 2.8$, $c = 5$, and $K = 0.9$, as Fig. 3.5 shows.

There are other ways of using the precursor model for analysing sticking probability curves, but they do not introduce any important new principles. We shall therefore next turn to an outline of the experimental methods used to obtain these curves.

5. Experimental methods for obtaining sticking probability curves

The fundamental requirements of an apparatus suitable for measuring sticking probability curves can be deduced by recourse to the kinetic theory of gases. We shall start by using this theory to make an order-of-magnitude estimate of a convenient experimental pressure of reactive gas.

The collision rate R of a gas of molecular weight m at temperature T and pressure p is

$$R = \frac{p}{(2\pi mkT)^{1/2}} \quad \text{collisions cm}^{-2}\,\text{s}^{-1}.$$

The unit most commonly used for pressure is the Torr (1 Torr = 1 cm Hg = 133.3 N m^{-2}). At 298 K the collision rate is

$$R_{298} = \frac{p}{m^{1/2}} \times 2 \times 10^{21} \quad \text{collisions cm}^{-2}\,\text{s}^{-1}\,\text{Torr}^{-1}.$$

For a gas of mass 28

$$R_{298} = 3.8 \times 10^{20}\, p \quad \text{collisions cm}^{-2}\,\text{s}^{-1}\,\text{Torr}^{-1}.$$

Now a typical metal surface has about 10^{15} surface sites cm^{-2}. Therefore if, say, half the collisions lead to adsorption, a fairly common result at low coverages, the rate of adsorption is

$$R_{ad} = 1.9 \times 10^{20}\, p \quad \text{molecules cm}^{-2}\,\text{s}^{-1}\,\text{Torr}^{-1}$$

$$= 1.9 \times \frac{10^{20}}{10^{15}}\, p = 1.9 \times 10^5\, p \quad \text{surface layers s}^{-1}\,\text{Torr}^{-1}.$$

It is convenient for the time of an experiment to be not less than 100 s. Thus we require that

$$1.9 \times 10^5 p \times 100 < 1,$$

i.e. for such a system the working pressure p would be less than about 5×10^{-8} Torr. In practice, sticking probabilities vary considerably from system to system so that the pressure range 10^{-8}–10^{-6} Torr is common.

The residual gases in a well-pumped vacuum chamber often include carbon monoxide and hydrogen, both of which are reactive towards many metals. It is necessary, therefore, to reduce the pressure in the reaction chamber to a small fraction of the working pressures to avoid significant interference. Background pressures in the low 10^{-10} Torr range are required. The methods for production and measurement of these ultrahigh vacua are well established and descriptions can be found in books on vacuum techniques.

Having identified the appropriate experimental conditions for the gas we now need to consider the surface. As mounted in an ultrahigh vacuum (UHV) chamber the sample will be covered with a layer of atmospheric gas and, possibly, impurities from fabrication. It must therefore be cleaned prior to admission of the gas whose adsorptive behaviour is under investigation. Methods for doing this are discussed next.

Methods of cleaning surfaces

The generation of a clean surface may involve either starting with an existing contaminated surface and removing the contaminant, or forming a new surface *in situ*. These methods will be considered in turn.

1. Removal of surface contaminants. There are two widely used methods for removing surface contamination and often samples are cleaned by applying these consecutively. The first method is by chemical cleaning. A common surface impurity is carbon and this may be removable as CO and CO_2 by heating the sample in oxygen. Typically, exposure to oxygen at 10^{-6} Torr for several hours with the sample at 1000–1500 K, perhaps on more than one occasion, is required. The resulting layer of oxygen is removed either by heating the sample *in vacuo* or by reduction with hydrogen. The former is preferred when it is possible.

This method also works well for other non-metallic impurities (e.g. boron, sulphur, and phosphorus) and is widely used for the high melting transition metals whose oxides are less stable than the metal. Examples include tungsten, molybdenum, rhenium and tantalum. The second method is suitable for metals from which a surface layer of oxygen cannot readily be removed or where a surface impurity is present which is not susceptible to chemical cleaning. It requires a beam of argon ions, which may be produced by electron bombardment, the ions then being accelerated through, typically, 600 V. These energetic ions eject atoms from the surface, a process called 'sputtering'. Both surface metal and impurity atoms are sputtered. The resulting pock-marked surface is prepared for reaction by annealing *in vacuo*.

2. Generation of a new surface. A metal film can be formed on the walls of the UHV chamber if the sample is heated sufficiently strongly. Films of metals which melt before evaporating can be prepared from a bead of metal supported on a hairpin of tungsten and others can be prepared by direct evaporation from a filament. The resulting clean surface consists of many tiny crystallites and exposes a wide variety of

crystal planes. Before admitting gas, the surface is often stabilized by warming to a temperature well above that to be used for adsorption (provided that the material of the UHV chamber, which is usually Pyrex glass, allows this).

Although not used for sticking probability measurements, one other method of cleaning surfaces is worth mentioning. This method is often used to clean tips for field emission experiments (to be described in Chapter 6). The sample is in the form of a needle-sharp point to which a high electric field is applied. Under extreme conditions of electric field at the the tip (about 4×10^8 V cm^{-1}) surface atoms, both impurity and substrate, are torn from the surface; this process is called 'field desorption'. The resulting clean surface is annealed before use.

There is a significant difference between typical specimens prepared by methods (1) and (2) (excluding field emission tips). Method (1) is particularly suitable for metal wires, ribbons, or single-crystal planes of transition metals, and the specimens typically have an area of about 1 cm^2. Such samples are usually mounted on electrical leads and are heated either by passing electric current through them or by electron bombardment from an adjacent incandescent tungsten cathode. For experiments involving the adsorption of simple reactive gases adsorbed on a transition metal, heating *in vacuo* to 1000–2000 K is often sufficient to remove the adsorbed layer. The clean metal surface is thereby regenerated and adsorption can be repeated. On the other hand a film evaporated onto a UHV chamber wall, whilst having the convenience of a much larger surface area, cannot usually be heated to a temperature sufficient to desorb an adsorbed layer before the glass chamber softens. It is thus necessary to lay down a new layer before repeating the experiment. This is a much more troublesome procedure than electrical heating. Results based on the former method are more numerous and more readily interpreted. They will be described next.

Adsorption by wires, ribbons, and single crystals

There is one further important factor in the design of the apparatus to be noted. A typical UHV chamber designed for measuring sticking probability curves might have a volume of about 2 l. Thus at a gas pressure of 1×10^{-7} Torr, the number of molecules in the chamber, calculated from kinetic theory, is about 10^{13}. For a sample of area 1 cm^2 this number is much less than that required for a monolayer, which may be up to 10^{15} molecules. It is therefore necessary to provide a flow of gas into the sample chamber. It has also been found advantageous for some experiments to maintain a connection between the sample chamber and the pumping system. In schematic form the basic features of an apparatus

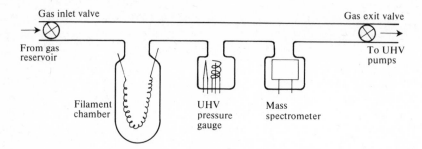

Fig. 3.6. Schematic diagram of the UHV apparatus for sticking probability curves. The pressure records are made using the UHV gauge or the mass spectrometer as appropriate.

designed specifically for the measurement of sticking probability curves are illustrated in Fig. 3.6. It is usual to pass a stream of gas from a reservoir through a valve to the UHV chamber, over the sample, and out through a valve to the pump. The valves are adjusted to produce a rate of flow of gas and an equilibrium pressure suitable for the particular system under scrutiny. There are two ways in which the sticking probability curve can be obtained.

Method 1

The UHV chamber is pumped out and the sample is cleaned. The gas inlet and outlet valves are then set to positions previously determined as suitable by trial and error. At this stage the system is in equilibrium, the surface being saturated with gas and the rate of admission of gas from the reservoir equalling the rate at which it is pumped away through the outlet valve. The sample is now heated to a temperature sufficient to remove the adsorbed layer (1000–2000 K, depending on the strength of adsorption). The gas evolved from the sample, which produces a rise in pressure, pumps away through the outlet valve and the pressure returns to its equilibrium value. At this stage the hot sample is neither adsorbing nor desorbing gas. On cooling adsorption begins, the effect being to produce an additional pump for the reactive gas. The pressure therefore falls. Eventually the surface becomes fully covered, the sticking probability returns to zero, and the gas regains its equilibrium pressure. The partial pressure of adsorbing gas during this sequence of events is recorded mass spectromet-

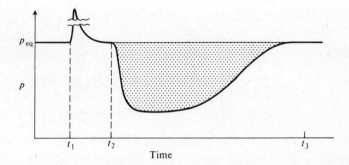

Fig. 3.7. Typical shape of the gas adsorption pressure record. At t_1 the sample is heated to clean it; at t_2 the heating supply is switched off and adsorption starts; at t_3 adsorption is complete and the pressure has returned to the equilibrium value.

rically. A typical adsorption curve is shown in Fig. 3.7. The results are analysed as follows.

The rate of change of pressure with time at any particular point on the adsorption curve can be written as

$$\frac{dp}{dt} = \left(\frac{dp}{dt}\right)_{pump} + \left(\frac{dp}{dt}\right)_{inlet} + \left(\frac{dp}{dt}\right)_{sample}. \qquad (3.8)$$

Let us consider each of the terms on the right-hand side in turn.

1. $(dp/dt)_{pump}$ is the pumping rate through the aperture connecting the UHV chamber to the pump. This rate is measured from the results of a separate experiment, in which the sample is inactive, as follows. A suitable pressure of gas ($\sim 10^{-6}$–10^{-7} Torr) is established in the UHV chamber and the inlet valve is closed rapidly. The gas then pumps away and the pressure versus time record has the form shown in Fig. 3.8. The

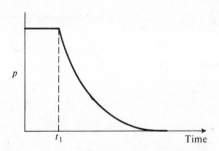

Fig. 3.8. Pressure–time record for measuring pumping constants. The gas admission valve is rapidly shut at t_1.

first-order decay curve is analysed using the equation

$$\frac{dp}{dt} = -kp.$$

where k is the pumping constant of the aperture, measured in units of time^{-1}. It is related to the pumping speed S of the aperture by $k = S/V$, where V is the volume of the apparatus. Thus in eqn (3.8) the term $(dp/dt)_{pump}$ is given by $-kP$.

2. $(dp/dt)_{inlet}$ is the rate at which gas is admitted to the UHV chamber. This is constant, since there is a high enough pressure (typically a few Torr) of gas in the reservoir that no measurable change occurs during an experiment.

When the sample is saturated and the system is at its equilibrium pressure p_{eq}, the rate of admission of gas equals the rate at which it is pumped. This latter is kp_{eq}. Therefore

$$\left(\frac{dp}{dt}\right)_{inlet} = kp_{eq}.$$

3. $(dp/dt)_{sample}$ measures the effect of adsorption on the pressure. The number n_g of gas molecules in the UHV chamber, whose volume is V l, is cpV (where c is a constant and is equal to 3.3×10^{16} l^{-1} Torr^{-1} at 298 K). Mass balance requires that the rate of adsorption equals the rate of loss of gas phase molecules, i.e.

$$\frac{dn}{dt} = -\frac{dn_g}{dt} = -cV\left(\frac{dp}{dt}\right)_{sample}$$

where n is the number of surface molecules. For a sample of area A the rate of adsorption is also given by

$$\frac{dn}{dt} = sApZ$$

where Z is the collision rate of the gas at a pressure of 1 Torr with unit area ($Z = 2 \times 10^{21}/m^{1/2}$ cm^{-2} s^{-1} Torr^{-1}). Thus

$$\left(\frac{dp}{dt}\right)_{sample} = -\frac{sApZ}{cV}$$

In the flat valley region of the pressure versus time curve, $dp/dt = 0$ and the sticking probability is constant at its initial value s_0. In this region therefore

$$s_0 = \frac{cVK}{AZ}\left(\frac{P_{eq}}{P} - 1\right) \tag{3.9}$$

The initial sticking probability is thus readily calculated once the experimental parameters V, A, and k have been measured.

Provided that dp/dt is much less than the terms on the right-hand side of eqn (3.8), and this is usually the case, eqn (3.9) can also be used to calculate values of the sticking probability s later on in the adsorption.

A modification of this experimental procedure is often used for gases which are decomposed by the heated sample. A particular example is hydrogen, which is usually atomized by metals at high temperatures ($T > 1300$ K). To overcome the difficulty this creates, the sample is heated *in vacuo* to clean it and cooled, and then the gas is let in. An example of this precedure for the adsorption of hydrogen and deuterium by a metal is shown in Fig. 3.9.

Curves such as those shown in Figs. 3.7 and 3.9 can also be used to calculate the uptake as a function of adsorption time. Since

$$\left(\frac{dn}{dt}\right) = -\left(\frac{dn}{dt}\right)_{gas} = spAZ,$$

$$n_t = \int_0^t spAZ \, dt$$

and substituting eqn (3.9) in its general form (i.e. substituting s for s_0) gives

$$n_t = cVk \int_0^t (p_{eq} - p) \, dt.$$

The integral does not have an analytical form but is evaluated graphically from the pressure record. It is the area between the p_{eq} line and the

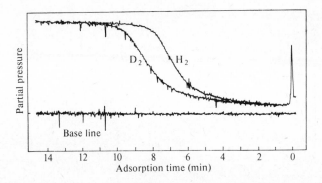

Fig. 3.9. Adsorption traces measured mass spectrometrically for the adsorption of hydrogen or deuterium on tungsten (polycrystalline). (From Gasser, Morton, Overton, and Szczepura 1971.)

pressure trace, as shown cross-hatched on Fig. 3.7. The sticking probability curve (s/s_0 versus θ) is then constructed by calculating s and θ at intervals of time between t_2 and the time at which s becomes immeasurably small.

Method 2

Another method of measuring sticking probability curves is to follow the first stages of the procedures already described, but instead of allowing adsorption to go to completion the partial layer is removed by rapidly heating ('flashing') the sample. The burst of gas thus produced is recorded. The experiment is then repeated, the surface layer being flashed off after a different time. Numerous repetitions thus provide a record of the pressure burst as a function of adsorption time.

The amount of gas adsorbed can be calculated from the pressure burst in a number of ways, depending on the experimental conditions.

1. If the gas is evolved rapidly compared with the rate of pumping by the aperture, an insignificant amount is lost during the flash, as shown in Fig. 3.10(a). Then

$$-\left(\frac{dn}{dt}\right) = \left(\frac{dn}{dt}\right)_{gas} = cV\frac{dp}{dt} - n = cV\int_{p_i}^{p_f} dp = cV\,\Delta p \qquad (3.10)$$

where p_i is the initial pressure, p_f is the maximum pressure and c is the constant in eqn (3.9). The pressure rise Δp thus gives the coverage n.

2. When the pumping rate is not slow compared with the rate of desorption from the surface an appreciable amount of gas is pumped away. The pressure burst now has the shape shown in Fig. 3.10(b). Then, as before (eqn (3.8)),

$$\frac{dp}{dt} = \left(\frac{dp}{dt}\right)_{pump} + \left(\frac{dp}{dt}\right)_{inlet} + \left(\frac{dp}{dt}\right)_{sample}$$

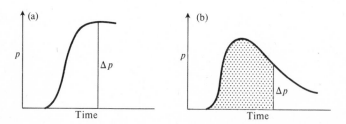

Fig. 3.10. Pressure records of gas evolution in the flash filament experiment: (a) negligible pumping rate; (b) appreciable pumping rate. In (a) the uptake is calculated from Δp; in (b) it is calculated from the cross-hatched area plus Δp.

Therefore

$$\left(\frac{dp}{dt}\right)_{\text{sample}} = \frac{dp}{dt} + kp - kp_{\text{eq}}$$

$$\Delta p_{\text{sample}} = \Delta p + k\int_0^t (p - p_{\text{eq}})\, dt.$$

Thus the uptake by the surface is calculated from the pressure rise Δp and the area under the desorption curve. Naturally the time after the flash at which the area is calculated must be longer than the time required for all the gas to desorb; the criterion is that the pressure record must have entered the region of exponential decay. A suitable time is shown in Fig. 3.10(b). Again Δp_{sample} is multiplied by the factor cV to obtain $(n_{\text{surface}})_t$. Alternatively, the total area can be measured if the entire desorption pressure burst is recorded. Next a graph of $(n_{\text{surf}})_t$ is plotted against t, the slope of the graph at any point being the rate of adsorption. An example of such a graph for hydrogen and oxygen on tungsten is shown in Fig. 3.11. The rate of collision of gas molecules is AZp, which is obtained from the pressure versus time record as before. Then

$$s = \frac{dn/dt}{AZp}.$$

Thus a sticking probability curve is obtained.

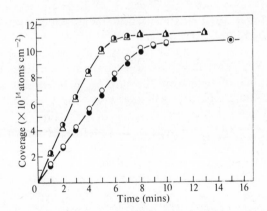

Fig. 3.11. The dependence of the uptake of the hydrogen isotopes on the time of adsorption for two tungsten filaments: upper curve, $(p_{\text{eq}})_{D_2} = 2^{1/2}(p_{\text{eq}})_{H_2} = 5 \times 10^{-8}$ Torr ($\bullet$, deuterium; $\triangle$, hydrogen); lower curve, $(p_{\text{eq}})_{D_2} = 2^{1/2}(p_{\text{eq}})_{H_2} = 6.6 \times 10^{-8}$ Torr ($\bullet$, deuterium; $\bigcirc$, hydrogen). Each filament behaves in the same way towards the isotopes, thus indicating the absence of a kinetic isotope effect in adsorption. (From Gasser *et al.* 1971.)

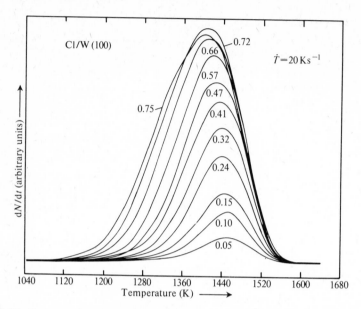

Fig. 3.12. Desorption of chlorine atoms from tungsten (100). The mass spectrometer was in line of sight from the sample. The curves show the essentially temperature-independent maximum characteristic of first order desorption, eqn (3.14). The curves yield a coverage-independent energy of desorption of 82 kcal mol^{-1}. (From Kramer and Bauer 1981.)

This method was used in the earliest measurements and has subsequently been found to be particularly useful for observing the desorption of species which condense on or react with the walls of the vessel. For these latter experiments the mass spectrometer ionization region is arranged to be in line of sight from the surface. A fraction of the desorbed species is thus ionized before colliding with a wall, and the mass spectrometer trace reflects the effective partial pressure of the desorbed molecules. An example of the use of this method is shown in Fig. 3.12. There is, however, one problem when very reactive species are being desorbed and that is how to obtain a reliable value for the absolute magnitude of the uptake. The fractional coverage, expressed as

$$\theta = n/n_{max},$$

is readily available, but n_{max} may have to be estimated by separate calibration or by making assumptions as to the adsorptive capacity of the surface.

In the methods described above, the gas molecules impinge on the surface with the usual random angular distribution. A number of experiments have been performed in which a directional character is imparted to the gas flow, leading to collisions with a defined part of the adsorbent. Such experiments yield more accurate values of sticking probabilities and will be dealt with next.

Directional flow methods

Molecular beam source

The basis of this experiment is the formation of the gas into a molecular beam by passing it through a series of collimating orifices (King and Wells 1972). The beam, which has an area of about $0.04 \, \text{cm}^2$, then collides with the crystal surface which has an area greater than $0.5 \, \text{cm}^2$. Thus, only a small central part of the target becomes covered with gas. The flux in the beam is typically 10^{12}–10^{13} molecules $\text{cm}^{-2} \, \text{s}^{-1}$, giving equivalent pressures at the target similar to those in the experiments described earlier. The apparatus is shown schematically in Fig. 3.13.

Sticking probability curves are obtained by first recording the pressure versus time variation in the filament chamber (whose volume is about $1 \, \text{l}$) when the crystal is in the path of the beam. The experiment is repeated after the crystal has been moved to an out-of-beam position so that adsorption only occurs after gas molecules have collided with the walls. The shapes of the two p versus t curves are shown in Fig. 3.14. Sticking probabilities are calculated by selecting times at which the exposures, i.e. $\int_0^t p \, dt$, are the same in the two experiments. These are the areas shown cross-hatched in Fig. 3.14. Then the sticking probability is given by

$$s = \frac{p_2 - p_1}{p_2}$$

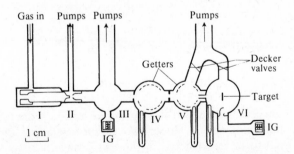

Fig. 3.13. Diagrammàtic representation of the molecular beam apparatus for sticking probability determination (approximately to scale). The molecular beam is generated in sections I–IV. (From King and Wells 1972.)

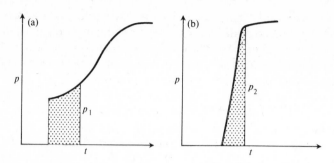

Fig. 3.14. Pressure versus time records for molecular beam adsorption studies of nitrogen on tungsten: (a) crystal in path of beam; (b) crystal out of beam. (From King and Wells 1972.)

where the pressures p_1 and p_2 are those recorded at time t for the in-beam and out-of-beam experiments respectively. The relevant coverage is obtained from $\int_0^t p \, dt$ of the in-beam experiment in a manner analogous to that discussed already.

An alternative method of detecting the reflected molecules is to use a field emission tip situated opposite the crystal face on which the molecules impinge.

Effusive source

An effusive source has also been used for precision measurements of sticking probabilities (Madey 1973). The crystal was positioned opposite the precision gas dosing system which produced a highly stable flux of effusing molecules. A known fraction f of these molecules collided with the crystal. For a UHV chamber with very rapid pumping, the sticking probability is related to the pressures (measured mass spectrometrically) by

$$s = \frac{p_{eq} - p}{f p_{eq}}.$$

The coverage at any time $n(t)$ is obtained from

$$n(t) = (F + F_r) \int_0^t s \, dt$$

where F is the flux of molecules from the source and F_r is the random flux from background molecules. These two equations allow the sticking probability curve to be constructed in the usual way.

Mention has already been made of flash desorption for determining uptakes at chosen stages of adsorption. However, it often transpires that desorption from a layer occurs in stages, so that a number of pressure peaks is observed. Thus a desorption 'spectrum' is produced. The nomenclature of such peaks is by no means standard, but it is common to refer to peaks produced at temperatures below room temperature as γ peaks, those produced at or a little above room temperature as α peaks, and those produced at high temperatures as β peaks. When the peaks are themselves composite, subscript numerals are used, e.g. β_1, β_2. This procedure will be followed hereafter. Desorption peaks may not only give information about the numbers of molecules on the surface but also allow the kinetic order of the desorption process and its activation energy to be calculated.

Adsorption by films

Films are prepared by evaporation of the metal onto the walls of the UHV chamber and differ from the suspended samples previously discussed in two important ways: (i) the geometric area exposed to the gas is much greater and (ii) the film is made up of microcrystallites and is therefore rough. Of course, polycrystalline wires or ribbons are not necessarily completely smooth on the atomic scale, but whereas the roughness factor (the ratio of true area to the geometric area) probably does not exceed 1.5 for such samples, a factor of 5 might be typical for a film. The consequences of (i) and (ii) are that a different design of apparatus is required and that care must be exercised in interpreting the results. The flow technique is modified by omission of the aperture to the pumps, gas being admitted to a closed-off chamber. The geometry of the apparatus is critical to the validity of the results obtained, since it is imperative that the film be uniformly exposed to the gas. This is achieved by using a spherical adsorption vessel with a gas source at the centre of the sphere (Hayward, King, and Tompkins 1967). The source is a small uniformly perforated glass bulb connected to the gas supply, the 'diffuser'. The arrangement is shown schematically in Fig. 3.15. A movable glass disc can be transferred to a position between the diffuser and the side-arm leading to the pressure gauge. With the disc in place, gas molecules enter the gauge after collision with the film; the pressure p_d is then due to rebounding molecules. With the disc removed, gas molecules enter the gauge directly from the diffuser. The gauge then experiences the same pressure as the film. Then

$$s = \frac{\text{no. of molecules adsorbing}}{\text{no. of molecules colliding}}$$

$$= \frac{\text{total no. of collisions} - \text{no. of molecules rebounding}}{\text{total no. of collisions}}$$

$$s = \frac{p - p_d}{p}.$$

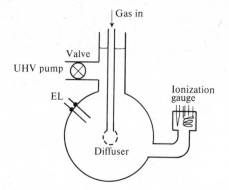

Fig. 3.15. Schematic diagram of the arrangement for dosing a metal film evaporated on the walls of the sphere. The film is deposited by withdrawing the diffuser and heating a hairpin of metal attached to the electrical leads EL. During this process the side-arm to the gauge is sealed off with a movable disc.

The sticking probability thus recorded is larger than that on a filament or ribbon of the same metal. The reason is that the roughness of the film allows an incoming molecule to undergo several collisions before returning to the gas phase. The true sticking probability can therefore only be obtained by making assumptions as to the average number of times a molecule incident from the gas phase is able to collide with the rough surface before escaping back to the gas. If this number is $\langle c \rangle$ then

$$\frac{s}{s_{app}} = \{(1 - \langle c \rangle)(1 - s)\}^{-1}$$

where s_{app} is the apparent sticking probability.

Desorption

In general the rate of desorption from a layer can be described by the Arrhenius equation in the form

$$-\frac{dn}{dt} = \nu n^a \exp\left(\frac{-E_d}{RT}\right) \tag{3.11}$$

where n is the surface coverage (in particles cm^{-2}), a is the kinetic order, E_d is the energy for desorption and ν is the frequency factor. The usual method of obtaining ν, a, and E_d is to heat the sample (often in a few seconds) and record n as a function of time. Mathematical treatment of the results is facilitated if there is a functional relationship between the rise $(T - T_0)$ in the temperature of the sample and the time of heating t.

The two most convenient and common heating programmes are

$$T = T_0 + \beta t \quad \text{(linear)} \tag{3.12}$$

and

$$\frac{1}{T} = \frac{1}{T_0} + bt \quad \text{(hyperbolic)}$$

It is usually easier to obtain the surface coverage during a flash from the gas phase record (since $dn_{gas}/dt = -dn/dt$) rather than by a direct measurement of the surface layer itself, because rapid pressure changes can be recorded readily; however, this is not true of surface properties sensitive to adsorbed layers. We shall now consider how the pressure record can be analysed when a linear heating programme is used.

1. Analysis of desorption curves

The main factors which contribute to the pressure record of a desorption are similar to those that affect adsorption and can be listed as follows.

1. The gas being admitted to the system at a constant rate kp_{eq} (see eqn 3.8).
2. Pumping of gas through the aperture to the pumps at a rate kp (see eqn 3.8).
3. Evolution of gas from the sample at a rate (in molecules s^{-1}) given by $A\,dn/dt$ for a sample of area A.
4. Miscellaneous factors, such as possible effects of the walls and interference by the pressure-measuring device. It is usual to arrange the experimental conditions to minimize these undesired contributions, but wall effects in particular may limit the accuracy of the results. These wall effects arise from physical interactions between gas and surface and are especially troublesome when polar gases are under investigation.

Ignoring 4 we can write

$$\frac{dp}{dt} = k(p_{eq} - p) - \frac{A}{cV}\frac{dn}{dt}.$$

The analysis of desorption curves is facilitated if either of two inequalities is obeyed: rate of pumping ≪ rate of evolution of gas, i.e. slow pumping, *or* rate of pumping ≫ rate of evolution of gas, i.e. fast pumping. We shall consider these in turn.

Slow pumping

$$\frac{dp}{dt} \gg k(p_{eq} - p)$$

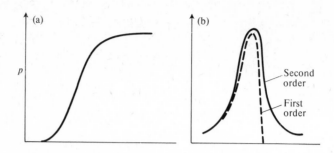

Fig. 3.16. Pressure versus time records during flash desorption; (a) slow pumping; (b) fast pumping, for first- and second-order desorption kinetics.

so that

$$\frac{dp}{dt} = -\frac{A}{cV}\frac{dn}{dt}.$$

The general form of the pressure–time record is shown in Fig. 3.16(a). Since a linear heating programme is under discussion the ordinate is linear in temperature as well as in time. This fact is made use of in converting the general rate (eqn (3.11)) which has the two variables t and T into an equation with just T as the variable.

$$T = T_0 + \beta t$$

Therefore

$$\frac{dT}{dt} = \beta$$

$$\frac{dn}{dt} = \beta\frac{dn}{dT}.$$

Thus

$$-\frac{dn}{dT} = \frac{\nu}{\beta}n^a\exp\left(\frac{-E_d}{RT}\right). \tag{3.13}$$

The initial uptake n_0 can be calculated from eqn (3.10) and the uptake at any time during the flash from the difference between n_0 and the amount of gas in the gas phase, also calculated from eqn (3.10), using the instantaneous value of Δp. The slope of the desorption curve yields dn/dT. Hence a plot of $\ln\{(1/n^a)(dn/dT)\}$ versus $1/T$ is linear for $a = 1$ when the reaction is first order in surface concentration or linear for $a = 2$

when the reaction is second order. The slope of the line gives E_d. For a known value of β (the heating rate) and the established values of a and E_d it is then possible to calculate the pre-exponential factor ν.

Fast pumping

$$\frac{dp}{dt} \ll k(p_{eq} - p)$$

so that

$$\frac{dn}{dt} = \frac{cV}{A} k(p_{eq} - p).$$

In this case therefore the desorption rate is directly proportional to the pressure. However, because of the difficulty of measuring rapid pumping constants accurately, it is usual to analyse the results somewhat differently from the way used in the slow pumping regime. The first consideration is the shape of the desorption curve. For rapid pumping the gas is evolved as a pressure transient i.e. the pressure versus t (or T) record is a peak. The shape of this peak depends upon the kinetics of desorption. Calculated curves for first-order and second-order desorptions are shown in Fig. 3.16(b). The most significant feature of these curves is the symmetry of the second-order curve about the peak temperature T_p.

The next stage is to determine the maximum rate of desorption. When this rate has been reached, d^2n/dt^2 and, in the linear heating programme, d^2n/dT^2 are both zero. Thus if the general rate equation (3.13).

$$-\frac{dn}{dT} = \frac{\nu}{\beta} n^a \exp\left(\frac{-E_d}{RT}\right)$$

is differentiated with respect to temperature

$$-\frac{d^2n}{dT^2} = \frac{\nu}{\beta} \left\{ an^{a-1} \frac{dn}{dT} \exp\left(\frac{-E_d}{RT}\right) + \frac{n^a E_d}{RT^2} \exp\left(\frac{-E_d}{RT}\right) \right\}.$$

At the peak in the pressure burst $d^2n/dT^2 = 0$, and the temperature is T_p. Therefore

$$an^{a-1} \frac{dn}{dT} = -\frac{n^a E_d}{RT_p^2}.$$

Substituting from eqn (3.13) for dn/dT gives

$$\frac{\nu a n^{a-1}}{\beta} \exp\left(\frac{-E_d}{RT_p}\right) = \frac{E_d}{RT_p^2}.$$

for first-order desorption, $a = 1$ and

$$\frac{E_d}{RT_p^2} = \frac{\nu}{\beta} \exp\left(\frac{-E_d}{RT_p}\right). \tag{3.14}$$

We thus come to the important conclusion that T_p is independent of the initial coverage for a first-order desorption. A series of desorption peaks at increasing coverage whose position does not change is indicative of first-order desorption kinetics. An example is shown in Fig. 3.12.

For a second-order desorption, $a = 2$ and

$$\frac{E_d}{RT_p^2} = \frac{2\nu n_p}{\beta} \exp\left(\frac{-E_d}{RT}\right)$$

where n_p is the coverage at T_p.

Rearranging gives

$$\ln\left(\frac{\beta E_d}{2R\nu n_p}\right) = 2 \ln T_p - \frac{E_d}{RT_p}. \tag{3.15}$$

As we have already noted that the second-order curve is symmetrical, it follows that $n_p = n_0/2$ where n_0 is the initial coverage. Thus we see from eqn (3.14) that T_p decreases as a_0 increases. So a series of desorption peaks at increasing coverage whose maxima shift to lower temperatures is indicative of second-order desorption kinetics. Unfortunately, a complexity arises when, as happens not infrequently, a first-order desorption has an activation energy which decreases as the coverage increases. When this happens the peak also shifts to lower temperatures at higher coverages. The distinction between first and second order then rests on a consideration of the shape of the desorption curve.

Equations (3.14) and (3.15) allow estimates of E_d to be made. For a first-order process E_d can be obtained directly if a value is assumed for ν. It is common to use $\nu = 10^{13} \text{ s}^{-1}$. The justification given for this assumption arises from the transition state theory of chemical reactions which has a pre-exponential factor $(kT/h) = 1.6 \times 10^{-13} \text{ s}$ at 300 K. Alternatively, it may be argued that vibrational frequences are about 10^{13} s^{-1} and that vibration of the surface bond is a prerequisite for a first-order desorption process. However, these lines of argument have been questioned and values of ν differing by several powers of 10 (higher and lower) have been observed. In any case, as long as the value of ν used to calculate E_d is quoted, no confusion arises.

Alternatively, and better when practicable, β can be varied. Taking logarithms of eqn (3.14) we have

$$\ln\frac{E_d}{R} - 2 \ln T_p = \ln \nu - \ln \beta - \frac{E_d}{RT_p}.$$

Differentiating with respect to $\ln T_p$ yields

$$-2 = -\frac{d(\ln \beta)}{d(\ln T_p)} + \frac{E_d}{RT_p}$$

or

$$\frac{d(\ln \beta)}{d(\ln T_p)} = 2 + \frac{E_d}{RT_p}$$

Thus by plotting a graph of $\ln \beta$ versus $\ln T_p$ and measuring the slope, a value of E_d is obtained. It is desirable to vary β by two orders of magnitude for reasonable accuracy.

For second-order processes the substitution $n_p = n_0/2$ is made in eqn (3.15), giving after rearrangement

$$\ln(n_0 T_p^2) = \frac{E_d}{RT_p} - \ln \frac{\nu R}{\beta E_d}.$$

Thus a plot of $\ln(n_0 T_p^2)$ against $1/T_p$, if linear, confirms the second-order kinetics and its slope yields E_d.

So far we have taken the simplest possible examples of desorptions which occur in a single pressure burst, without considering what influence, if any, the precursor states discussed in adsorption might have on the desorption curves. We discuss these matters next, while noting that other methods of analysis of desorption curves besides those given above are possible.

2. Some complexities in desorption

Multiple binding sites

We have already noted that a desorption spectrum may be produced when a layer is flashed. The order of increasing bond strength in the states is $\gamma < \alpha < \beta$. A much studied system which exhibits all these states and illustrates the way in which the results can be interpreted is the desorption of CO from tungsten.

Adsorption of CO on a polycrystalline tungsten wire at room temperature was followed by flashing and showed the build-up of adsorbed molecules in the strongly bonded β state, which exhibited fine structure. In the later stages of adsorption the sticking probability of this state declined and a weaker α state was filled. The relative desorption curves are shown in Fig. 3.17. The inferences drawn from this set of experiments were that CO was adsorbed non-dissociatively since the peak temperatures were independent of coverage. As we have just seen, this observation implies first-order desorption kinetics, provided that the energy for

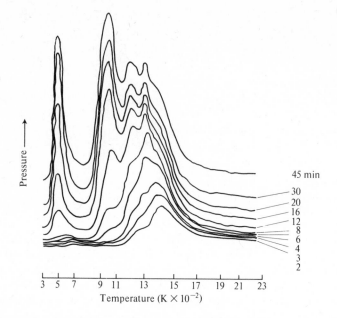

Fig. 3.17. Desorption of CO from tungsten (polycrystalline) after increasing exposure. (From Redhead 1961.)

desorption is constant, and thus a molecular rather than a dissociated state. Indeed, at one time it was thought that the evidence for molecular adsorption of CO on all metals was conclusive (Ford 1970). The reason suggested for the multiple peaks was a different form of bonding to the various planes exposed by a polycrystalline sample. The alternatives of single-site bonding for the α peak and bridge bonding for the β peaks, i.e.

$$
\begin{array}{ccc}
\overset{\displaystyle O}{\underset{\displaystyle W}{\overset{|}{\underset{|}{C}}}} \;(\alpha) & \qquad & \overset{\displaystyle O}{\underset{\displaystyle W \qquad W}{\overset{|}{C}}} \;(\beta)
\end{array}
$$

were suggested. Subsequently, it has been recognized that an alternative model in which β-CO absorbs dissociatively, but with lateral repulsive interactions between the absorbed species, would also account for the β desorption peaks (Goymour and King 1973). However, the α peak was judged to be molecular.

Later spectroscopic evidence has tended to support the view that when the strongest CO-to-metal surface bonds are formed, as in this case, the dissociation of CO is to be expected. More is said about this is in Chapter 6. As well as the α and β states, an additional molecular state is formed on some single-crystal tungsten planes at 77 K. Known picturesquely as a 'virgin' state, it converts to the β state on warming.

The observation of multiple desorption peaks is a common occurence, but as the above discussion shows inferences as to the nature of adsorbed state must be made with caution. It is usual to ally desorption data with the results of other experiments in drawing conclusions about surface states.

Influence of precursor states on desorption

In view of the well-established and considerable influence of precursor states on adsorption, the possibility that they may also have a perceptible effect on desorption peaks was for a long time curiously neglected. To illustrate the results of the comprehensive analysis now available (Gorte and Schmidt 1978; Cassuto and King 1981) we shall consider the treatment of a non-dissociative state formed via a precursor with a constant heat of adsorption. The potential energy diagram is illustrated in Fig. 3.18.

The kinetic scheme for non-reversible desorption is

$$A_{surface} \underset{k_a}{\overset{k_d}{\rightleftharpoons}} A^* \overset{k^*}{\longrightarrow} A_{gas}$$

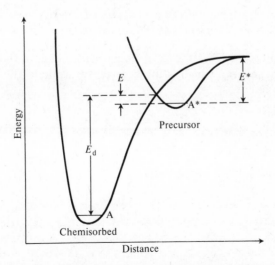

Fig. 3.18. Potential energy diagram for desorption including the precursor state. (From Gorte and Schmidt 1978.)

where the ks are velocity constants for the various reactants and A^* is the precursor state. Then at any coverage θ

$$\frac{d[A^*]}{dt} = k_d[A_s] - k_a[A^*](1-\theta) - k^*[A^*].$$

Applying the steady state approximation to $[A^*]$ gives

$$[A^*] = \frac{k_d[A_s]}{k^* + k_a(1-\theta)}$$

and the rate of desorption $R_d = k^*A^*$, which expressed in terms of fractional coverage θ becomes

$$R_d = \frac{k^*k_d\theta}{k^* + k_a(1-\theta)}.$$

Now two limiting cases can be considered:
1. $k^* \gg k_a$, when $R_d = k_d\theta$;
2. $k^* \ll k_a$, when

$$R_d = \frac{k^*k_d}{k_a}\frac{\theta}{1-\theta};$$

Case 1 is simply the normal first-order desorption process. In case 2 substitution for the velocity constants by the Arrhenius equation, i.e. $k = \nu \exp(-E/RT)$, yields

$$R_d = \frac{\nu^*\nu_d\theta}{\nu_a(1-\theta)} \exp - \left(\frac{E_d + E^* - E_a}{RT}\right).$$

In this expression $E_d + E^* - E_a$ is numerically equal to the difference in energy of A_s and A_g, which is the heat of adsorption. The equation can be rewritten as

$$-\frac{d\theta}{dt} = \frac{\nu'\theta}{1-\theta} \exp\left(\frac{-E_d}{RT}\right)$$

where $\nu' = \nu^*\nu_d/\nu_a$
    For the linear heating programme

$$\frac{d\theta}{dt} = \beta\frac{d\theta}{dT}.$$

Therefore

$$-\frac{d\theta}{dT} = \frac{\nu'}{\beta}\frac{\theta}{1-\theta} \exp\left(\frac{-E_d}{RT}\right).$$

Differentiating with respect to temperature and noting that at the peak maximum $d^2\theta/dT^2 = 0$ yields

$$\frac{E}{RT_p^2} = \frac{1}{(1-\theta_p)^2} \frac{\nu}{\beta} \exp\left(\frac{-E_d}{RT}\right). \quad (3.16)$$

Comparison of eqn (3.16), which is for the peak temperature of a first-order desorption when a precursor state is involved, with eqn (3.14), which is for direct desorption, shows that the difference is the term $(1-\theta_p)^{-2}$. Thus if desorption is carried out at low initial fractional coverages, the coverage at the peak θ_p will be small compared with unity

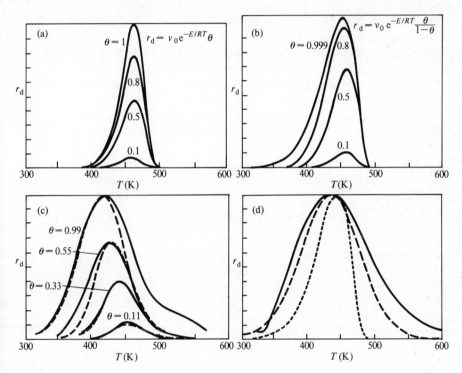

Fig. 3.19. (a) Normal first-order theoretical flash desorption curve; (b) first-order flash desorption curve calculated from eqn (3.16) (desorption parameters: $\nu_0 = 10^{13}\ \text{s}^{-1}$; $\beta = 100\ \text{K s}^{-1}$; $E_d = 109\ \text{kJ mol}^{-1}$). The theoretical and experimental desorption curves of CO from platinum are compared in (c) and (d). A more complex model was chosen than that used for (a) and (b) which gave a coverage dependence of the form $\theta^2/(1-\theta)$. (c) ——, experimental curve; - - -, calculated for $\nu_0 = 10^{13}\ \text{s}^{-1}$, $\beta = 100\ \text{K s}^{-1}$ and $E_d = 100\ \text{kJ mol}^{-1}$. (d) ——, experimental curve; · · ·, calculated from eqn (3.14) for $\nu_0 = 10^{13}\ \text{s}^{-1}$ and $\beta = 100\ \text{K s}^{-1}$; - - -, calculated using $\theta^2/(1-\theta)$.

and the term $(1 - \theta_p)^{-2}$ will not introduce a major shift in the peak temperature or the consequential calculated value of E_d. Even at high coverages the effect is not large. The peaks for a typical first-order desorption, with and without the $(1 - \theta_p)^{-2}$ term, are compared in Figs. 3.19(a) and 3.19(b). As can be seen, the precursor has more effect on the shapes of desorption curves, which are significantly broadened near saturation, then on T_p, which shifts by only 12 K, or on E_d, which increases by only 3 kJ mol^{-1}.

Some comparisons with experiment for the desorption of CO from two single-crystal planes of platinum, the (110) and (111), are shown in Figs. 3.19(b) and 3.19(c). On the former plane the agreement, though not perfect, is still better than that reached by allowing E_d to decrease with increase in θ_{CO} (lower E_d leads to earlier desorption at higher θ_{CO} and thus to a broader peak). On the former plane the precursor theory provides an improvement on direct desorption, but still does not broaden the peak enough. In this case it is thought that E_d really does decrease at higher coverages and broaden the peak further.

Thus far we have considered experimental methods and related theories in which the behaviour of the surface layer is inferred from observations made on the gas phase, namely the time dependence of the partial pressure. We now turn to a consideration of some techniques which respond to a property of the surface and to the way in which this property is modified by adsorption. No attempt will be made to give an exhaustive account, but a selection of those experiments most frequently encountered will be made.

4. Metallic structures and bonding

Introduction

Experience has shown that the behaviour of metals in adsorption and catalysis often depends considerably on the particular surface plane which is exposed to the gas. The most widely used technique for the study of this surface-specific reactivity is the diffraction of low energy electrons, i.e. electrons with energies in the range 20–500 V. The experiment consists of exposing the clean surface to a beam of electrons and measuring the position and intensity of the resulting diffracted beams. Subsequently, gas is admitted and is allowed to adsorb to a chosen coverage; the changes in the diffracted beams are recorded. In essence, this procedure represents a development of the classic experiment of Davisson and Germer in which the wave-like properties of electrons were demonstrated. An account of the interpretation of diffraction experiments forms the subject of the next chapter. However, to understand the results, some acquaintance with metallic structures and bonding is required together with an appreciation of the diffraction process. These will be the concern of this chapter.

Metallic structures

To describe the structures of metals, a simple model is used in which the crystal is composed of spheres of equal size, held together by forces which are non-directional in character. There are three main structures encountered in metals, known as face-centred cubic (also called cubic close-packed), hexagonal close-packed, and body-centred cubic. They are usually known by their initials, f.c.c., h.c.p., and b.c.c. The first two of these, f.c.c. and h.c.c., can be constructed by first taking spheres and packing them as closely as possible in a single layer. The resulting structure is shown in Fig. 4.1(a). The next two layers are placed on top of the first, so that they pack as closely as possible. There are two ways in which this can be done.

1. By placing the second layer so that the atoms lie in the hollows between the atoms of the first layer, and then putting the third layer so that the atoms are immediately above the first layer. Thus an alteration which can be described as AB AB AB is produced. This gives the h.c.p. structure.

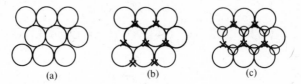

Fig. 4.1. Formation of close-packed metal structures: (a) close-packed layer A; (b) addition of second close-packed layer B (×). Placing the third layer above A gives the sequence ABAB .... which is the h.c.p. structure; (c) addition of a second close-packed layer B (×) followed by placing a third close-packed layer C (○) in an alternating position. The sequence is ABC ... which is the f.c.c. structure.

2. By placing the second layer in the hollows of the first layer as before but then placing the third layer in the alternative set of hollows which is not above the first layer. The fourth layer is now placed above the first. The sequence ABC ABC ABC is thus produced. This gives the f.c.c. structure.

The structures of the h.c.p. and f.c.c. lattices are shown in Figs. 4.2(a) and 4.2(b). The third structure, b.c.c., has an array of atoms as shown in Fig. 4.2(c). This structure leads to a slightly less dense packing of the atoms than is achieved in the other two (68 per cent of volume occupied compared with 74 per cent).

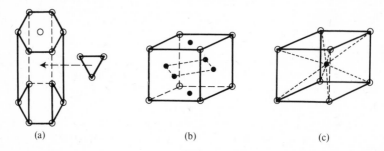

Fig. 4.2. Metallic structures: (a) hexagonal close packed; (b) face-centred cubic (○, corner atom; ●, centre atom); (c) body-centred cubic (○, corner atom; ●, centre atom).

Miller indices and surface structures

If we imagine taking a single crystal of a metal and making cuts at various angles through it, the surfaces thus exposed will have a variety of arrays of atoms. These are the single-crystal surfaces mentioned previously and in order to describe them a system of labelling is required. The labels

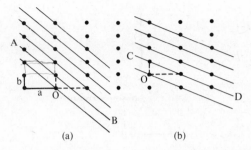

Fig. 4.3. Miller indices of planes in two dimensions. The set of planes parallel to AB are (2, 1); those parallel to CD are (1, 1).

usually encountered are the Miller indices and the method of deriving these will be described next.

Let us start by considering a two-dimensional array of lattice points (i.e. atoms) cut by a set of parallel planes (which are lines in two dimensions) chosen so that every lattice point lies on one of the planes. Two examples are illustrated in Fig. 4.3. If we now consider the set of planes in Fig. 4.3(a) and take as the origin the lattice point O, then the intercepts of the plane AB first encountered on the unit cell are $(a/2, b)$. The Miller indices are obtained by dividing the atomic spacing by the relative intercept, i.e. $(a/(a/2), b/b)$ or (2, 1). Since the choice of origin is arbitrary, all the parallel planes of this set have the same Miller index (2, 1). Similarly, the set of planes labelled (b) are designated as $(a/a, b/b) = (1, 1)$. A special situation arises when a set of planes is parallel to one of the axes. In this case the intercept on the axis is at infinity. The reciprocal of infinity is zero; so, for example, the horizontal planes through the lattice points are $(a/\infty, b/b) = (0, 1)$.

When a three-dimensional lattice is considered, a third index number is required. The plane is then designated (h, k, l), where $1/h$, $1/k$ and $1/l$ are the intercepts of the plane on the x axis, the y axis and the z axis respectively. The directions of the axes are chosen by convention as illustrated in Fig. 4.4; the plane of the paper is thus the yz plane. The Miller indices so defined serve to describe both a set of parallel planes within a lattice and the single-crystal surface resulting from cutting a crystal in such a way that its surface is parallel to the set. Let us take as an example a cubic lattice with atom spacings a. Then the planes parallel to the sides of the cube can be indexed by looking at Fig. 4.4. Thus a plane which includes the front face of the cube, drawn in full lines, does not cut the y or z axes, and cuts the x axis at a. This plane is indexed in the x direction as (a/a) and in the other two directions as (a/∞) and (a/∞); it is

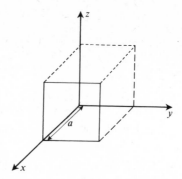

Fig. 4.4. Low Miller index planes of a cubic lattice. The front plane (———) is (100), the side plane (– – –) is (010), and the top plane (· · · ·) is (001).

therefore the (100) plane. Similarly the top plane is (001) and the side plane (010). Another important plane is that arising from a diagonal slice through the cube, as depicted in Fig. 4.5(a). This plane cuts each axis at a, so its Miller indices are $(a/a, a/a, a/a) = (111)$. Finally in the cubic system, let us consider the plane which divides the cube into two equal wedges, as shown in Fig. 4.5(b). This plane cuts the axes at (a, a, ∞) and is therefore the (110) plane.

In the h.c.p. system, two systems of nomenclature are encountered. The basal plane, which is the plane depicted in Fig. 4.1(a), is made up of an assemblage of equilateral triangles and can be conveniently described by

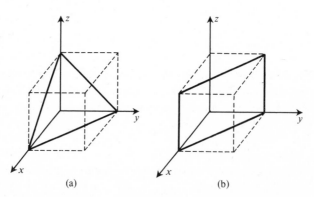

Fig. 4.5. Two diagonal planes of a cubic lattice (———): (a) (111) plane; (b) (110) plane.

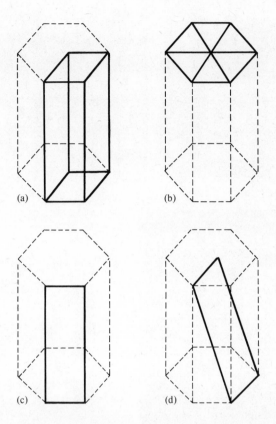

Fig. 4.6. (a) The unit cell for three-index notation of the h.c.p. lattice (——).
Three important planes of the h.c.p. lattice (——): (b) (0001) or (001) plane; (c)
(10$\bar{1}$0) or (100) plane; (d) (10$\bar{1}$1) or (101) plane.

three directions at 120°. A fourth direction perpendicular to this plane is
then required. The corresponding Miller indices are (h, k, i, l). However, a
three-index system is also used, which embodies the relationship

$$h + k = -i$$

Planes are indexed with reference to the axes of the unit cell shown in
Fig. 4.6(a). The three most frequently encountered h.c.p. planes are those
depicted in Figs. 4.6(b)–4.6(d).

The atomic arrangements of the planes described above are shown in
Fig. 4.7(a) for both cubic systems and the h.c.p. system.

bcc (100)

fcc (100)

hcp (001)

bcc (110)

fcc (111)

fcc (110)

Fig. 4.7(a)

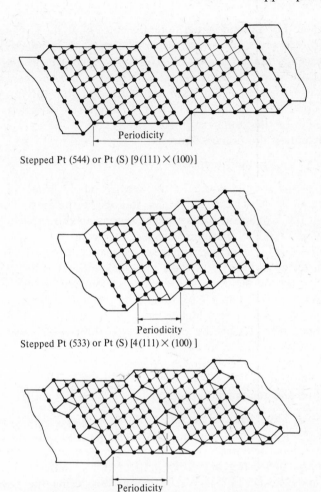

Stepped Pt (544) or Pt (S) [9 (111) × (100)]

Stepped Pt (533) or Pt (S) [4 (111) × (100)]

Kinked and stepped Pt (679) or Pt (S) [7 (111) × (310)]

Fig. 4.7. (a) Atomic arrangements of some frequently encountered metallic surface planes. Examples of metallic structures include: b.c.c. (tungsten, molybdenum), f.c.c. (platinum, palladium) and h.c.p. (rhenium). (Adapted from Nicholas 1965.) (b) Schematic representations of stepped and kinked planes. (From Somorjai 1977.)

Stepped planes

If a single crystal is cut in such a way that the surface plane is at an angle of a few degrees to one of the low index planes described above, the Miller indices involve high numbers. However this formal description of the plane may not do justice to the geometric interest of its atomic arrangement. For example, by a suitable choice of angle, a plane which is composed of a series of terraces several atoms wide connected by steps one atom high can be produced. This is in effect an atomic staircase whose tread width can be varied at will. A system of nomenclature for this kind of surface has come into use which describes the terrace width and the low index plane of which it is composed, together with the identification of the atomic step.

By convention the nomenclature is

$$\text{metal}(S(\text{tepped})) - n(h, k, l) \times (h', k', l')$$

where n is the atomic width of the terrace whose Miller indices are (h, k, l), and (h', k', l') are the Miller indices of the step. Let us now consider a few examples to illustrate this nomenclature (Nicholas 1965; Somorjai 1977). The platinum plane (544) consists of terraces nine atoms wide of (111) structure with steps consisting of rows of (100) atoms and is illustrated in Fig. 4.7(b). This plane is identified as $\text{Pt}(S) - 9(111) \times (100)$. Similarly the platinum (533) plane is identified as $\text{Pt}(S) - 4(111) \times (100)$ as illustrated in Fig. 4.7(b). If the slice through the crystal is chosen so that the step is part of a higher index plane, the result is to produce kinks in the terraces. Thus, for example, Pt(679) or $\text{Pt}(S) - 7(111) \times (310)$ is as illustrated in Fig. 4.7(b).

Real lattices and reciprocal lattices

Thus far we have been concerned with monatomic solids. However, adsorbed species are not, in general, monatomic, and in order to understand the results of diffraction experiments it is time to consider some more general properties of surface lattices (for a detailed account see Roberts and McKee 1978). We begin by considering the periodic arrangement of a surface. Starting at some arbitrary lattice point on a surface $\mathbf{r}$, consider a translation operation $\mathbf{T}$ being undertaken which involves two surface translation vectors $\mathbf{a}$ and $\mathbf{b}$. Then

$$\mathbf{T} = n_1 \mathbf{a} + n_2 \mathbf{b}$$

where n_1 and n_2 are integers. Now let us choose $\mathbf{a}$ and $\mathbf{b}$ so that the new lattice point $\mathbf{r}'$ given by

$$\mathbf{r} + \mathbf{T} = \mathbf{r} + n_1 \mathbf{a} + n_2 \mathbf{b}$$

is indistinguishable from $\mathbf{r}$. By letting n_1 and n_2 take all possible values, both positive and negative, the whole surface lattice can be constructed. This lattice has a repeating pattern defined by $\mathbf{a}$ and $\mathbf{b}$ and each unit of this pattern is known as the unit cell. However, in order to arrive at the

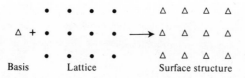

Fig. 4.8. Schematic representation of formation of the surface structure from the basis and the lattice.

full structure of the surface, the nature of the repeating unit located at each lattice point needs to be known. This repeating unit is known as the 'basis'. By combining the basis with the lattice the surface structure is reached. A schematic representation of this process is given in Fig. 4.8 where the basis is taken as a group of atoms in a triangular array and the lattice is rectangular.

It has been established that there are just five different types of surface lattice, known as the two-dimensional Bravais lattices. These are shown in Fig. 4.9. All of these except Fig. 4.9(e) are self-explanatory. This lattice can be looked on as centred rectangular, i.e. with $\mathbf{a} \neq \mathbf{b}$ and $\phi = 90°$. However, it can also be represented as an array of rhombuses. These alternatives bring us to the distinction between a unit cell and a primitive unit cell. The centred rectangle when repeated generates the surface and is therefore a unit cell, but it is not a primitive cell since this is defined as having lattice points only at its corners. Although there may be more than one way of constructing the primitive unit cell for a surface, by convention it is taken as one with the shortest translation vectors. Thus the rhombus drawn gives the primitive unit cell. However, the primitive unit cell does not do justice to the symmetry of the lattice and the centred cell is preferred.

As we shall see in the next section, the phenomenon of diffraction depends on the spacings between planes of atoms. A convenient method of depicting these spacings, and thus of understanding a diffraction pattern, is by way of the reciprocal lattice. The reciprocal lattice for a crystal is constructed from the three-dimensional solid lattice as follows (Levy 1968):

1. choose a point as origin;
2. from this point draw the normals to every set of planes in the real lattice;
3. make the length σ of the normal equal to the reciprocal of the interplanar spacing for the set of planes to which it relates;
4. place a point at the end of the normal.

The resulting array of points is the reciprocal lattice. If we call the

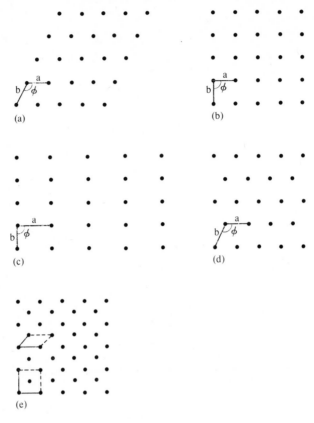

Fig. 4.9. Bravais lattices for a surface: (a) oblique, $\mathbf{a} \neq \mathbf{b}$, $\phi \neq 90°$; (b) square, $\mathbf{a} = \mathbf{b}$, $\phi = 90°$; (c) rectangular, $\mathbf{a} \neq \mathbf{b}$, $\phi = 90°$; (d) hexagonal, $\mathbf{a} = \mathbf{b}$, $\phi = 120°$; (e) rhombus or centred rectangular.

vectors in the real lattice defining the unit cell $\mathbf{a}$, $\mathbf{b}$, and $\mathbf{c}$ and the vectors in the reciprocal lattice $\mathbf{a}^*$, $\mathbf{b}^*$, and $\mathbf{c}^*$, (a) $\mathbf{a}^*$ is perpendicular to $\mathbf{b}$ and $\mathbf{c}$, and (b) $\mathbf{b}^*$ is perpendicular to $\mathbf{a}$ and $\mathbf{c}$, and (c) $\mathbf{c}^*$ is perpendicular to $\mathbf{a}$ and $\mathbf{b}$. A simple example in two dimensions will illustrate these statements. Consider two sets of planes of the real lattice at right angles and with spacings $\mathbf{a}$ and $\mathbf{b}$, as in Fig. 4.10. Application of the rules above produces two points in the reciprocal lattice, related to the origin O by $\mathbf{a}^*$ and $\mathbf{b}^*$. As can be seen $\mathbf{a}^*$ is perpendicular to $\mathbf{b}$ and $\mathbf{b}^*$ is perpendicular to $\mathbf{a}$.

In order to see how a reciprocal lattice is constructed, let us take the specific example of a surface with a square array of atoms whose lattice

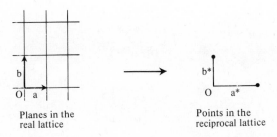

Planes in the
real lattice

Points in the
reciprocal lattice

Fig. 4.10. (a) Planes in the real lattice; (b) points in the reciprocal lattice.

constant is **a** and calculate the positions of the points of the reciprocal lattice (Levy 1968). As Fig. 4.7(a) shows, the frequently encountered f.c.c (100) and b.c.c (100) planes have just such a square structure. Following the rules above we proceed as follows:

1. the point O is chosen as origin;

2. noting that in two dimensions the equivalent of a plane is a line, we draw lines rather than planes through the lattice points (i.e. atoms) and consider a general line (h, k) such as the $(3, 2)$ illustrated in Fig. 4.11;

3. the perpendicular from O onto the line (h, k) is drawn.

We now need the length and direction of this perpendicular to plot the reciprocal lattice point. These essentially geometric quantities are calculated from the enlarged unit cell shown in Fig. 4.11 as follows.

Since

$$\frac{OB}{OA} = \frac{OC}{AC} \text{ (similar triangles)}$$

$$\frac{d}{a/h} = \frac{a/k}{(a^2/k^2 + a^2/h^2)^{1/2}}$$

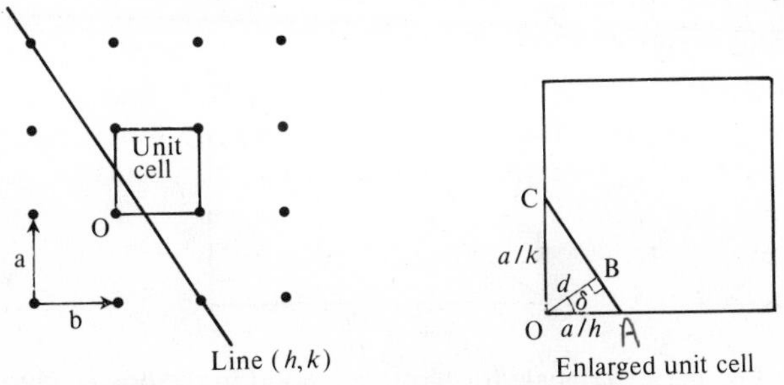

Line (h, k)

Enlarged unit cell

Fig. 4.11. Constructing a reciprocal lattice.

or

$$d = \frac{a}{(h^2 + k^2)^{1/2}}.$$

Therefore

$$\sigma = \frac{1}{d} = \left(\frac{h^2}{a^2} + \frac{k^2}{a^2}\right)^{1/2}.$$

The direction of σ is given by

$$\tan \delta = \frac{a/h}{a/k}$$

Now we construct a triangle in reciprocal space which fulfils the conditions that its hypotenuse is of length σ and its base angle is δ. Such a triangle is shown in Fig. 4.12. The point P is thus a point on the reciprocal lattice. Since h and k are integers, all points separated by a distance $1/a$ in the x direction and $1/a$ in the y direction of the reciprocal lattice satisfy the conditions for being points of the reciprocal lattice. That is to say, the reciprocal lattice of a real lattice of constant $\mathbf{a}$ is a square of lattice spacing $\mathbf{a}^* = 1/\mathbf{a}$. We may also note that a rectangular real lattice with lattice constants $\mathbf{a}$ and $\mathbf{b}$ gives a reciprocal lattice which is also a rectangle, with lattice constants $\mathbf{a}^* = 1/\mathbf{a}$ and $\mathbf{b}^* = 1/\mathbf{b}$, and that a real cubic lattice gives a reciprocal lattice which also cubic. It is described by the three integers h, k, and l, which are running indices in reciprocal space.

The phenomenon of diffraction can be treated in a variety of ways, and one which follows from the above discussion and which is well suited to interpreting the patterns observed in low energy electron diffraction (commonly abbreviated to LEED) is by way of the Ewald construction, which will be described next.

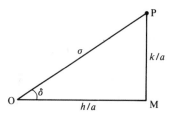

Fig. 4.12. Triangle for plotting a point in reciprocal space.

The Ewald construction

Let us first consider a two-dimensional crystal (not at this stage a surface plane) which has a square array of lattice points and thus a square array of reciprocal lattice points, as in Fig. 4.13(a). Now suppose that a beam of radiation of wavelength λ impinges on the crystal. The radiation can be represented by a vector **k** whose length is $1/\lambda$ and whose direction is the direction of the beam. To carry out the construction, one end of this vector is located at the origin O, already chosen for constructing the reciprocal lattice, and a circle of radius $1/\lambda$ is drawn with its centre at the other end A of the vector. If this circle passes through another reciprocal lattice point B, draw the triangle AOB and construct the perpendicular from A onto OB. Let us consider now the properties of triangle AOB.

  1. OB joins the origin to a reciprocal lattice point; thus OB is perpendicular to a set of planes in the real lattice whose interplanar spacing $d = 1/OB$.

  2. AC is perpendicular to OB and thus by definition is a lattice plane. Some of the planes parallel to AC and of spacing d are illustrated on the real lattice.

  3. The angle θ between AO and AC is the angle of incidence between the incoming radiation and the planes.

  4. In triangle AOC, $OC = (1/\lambda)\sin \theta$.

$$\sin \theta = \frac{OC}{K}$$
$$K = 1/T$$

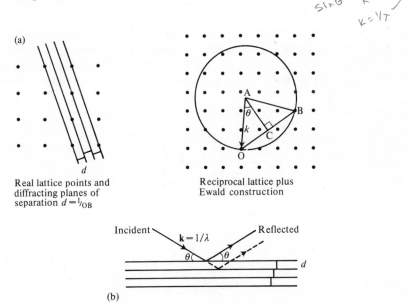

(a)

Real lattice points and diffracting planes of separation $d = 1/OB$

Reciprocal lattice plus Ewald construction

Incident · Reflected
$k = 1/\lambda$
(b)

Fig. 4.13. (a) The Ewald construction; (b) Bragg diffraction $n\lambda = 2d \sin \theta$.

$\frac{2}{\lambda} \sin \theta = \frac{1}{d}$

5. In triangle AOB, OB $= 2 \times$ OC $= 1/d$. Therefore

$$\lambda = 2d \sin \theta.$$

This equation is an example of the well-known Bragg condition for diffraction by a set of planes of spacing d, which is more conventionally drawn as shown in Fig. 4.13(b).

Some further points relating to diffraction can be deduced from Fig. 4.13.

First we note that there is no requirement that the circle should go through any points other than O on the reciprocal lattice. If it does not, this means that for radiation of this particular wavelength incident at this particular angle there is no set of planes which fulfil the Bragg condition. Secondly, if **k** is less than half the spacing in the reciprocal lattice it is impossible to draw a circle through any two points on the reciprocal lattice. This result implies (by substitution for **k** and a^*) that if $\lambda > 2a$ no diffraction is possible.

a?

Now let us consider specifically a surface plane on which a beam of monochromatic electrons impinges at normal incidence. This is the usual arrangement for LEED. This plane, a two-dimensional lattice, can be imagined as being formed by taking a normal three-dimensional lattice and stretching it along one of the directions of the unit cell, say **c**. The original reciprocal lattice, as we have already seen, is also a three-dimensional array of points, but as **c** becomes ever larger so **c*** becomes ever smaller. In the limit as **c** tends to infinity and the plane is formed, the **c*** reciprocal lattice points join together to form a line, known as a 'lattice rod' or 'lattice element'. The result is depicted schematically in Fig. 4.14.

In making the Ewald construction for a plane we start for simplicity's

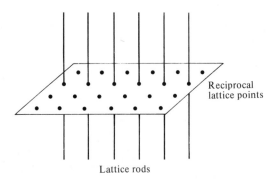

Lattice rods

Fig. 4.14. Schematic representation of lattice rods (for clarity one row only is shown). The reciprocal lattice points are in a plane perpendicular to the plane of the paper.

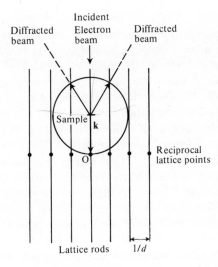

Fig. 4.15. Ewald construction in one direction for LEED.

sake with just a single row of reciprocal lattice points and the lattice rods through them, as illustrated in Fig. 4.15. The spacing between the reciprocal lattice points and rods is $1/d$, where d is, as before, the spacing in the real lattice. Next a vector of length $1/\lambda (= \mathbf{k})$, where λ is the de Broglie wavelength of the electrons $(\lambda = mv/h = (150/V)^{1/2})$, and in the same direction as that of the beam is drawn with one end at a point chosen as the origin O in the reciprocal lattice. The other end of the vector is chosen to be the origin in the real lattice, i.e. the sample. Again a circle of radius $1/\lambda$ is drawn, with its centre at the real lattice. If now vectors are drawn from the centre to the points of intersection of the circle with the lattice rods, these are the directions in which diffracted beams will be encountered. As illustrated there are just two equivalent diffracted beams, but if $\mathbf{k}$ is increased, i.e. the electrons are accelerated through a higher voltage, the radius of the circle increases. This may have two effects. First the angle at which the first diffracted beam is encountered decreases relative to the surface normal. Secondly, the circle may cut another set of lattice rods so that a second pair of beams may become observable.

What we have done so far is to identify the diffracted beams to be expected from a single set of lines drawn on a real two-dimensional lattice. When the full two-dimensional array is considered there are numerous sets of such lines to be drawn, with the corresponding sets of reciprocal points and lattice rods. Each time the Ewald construction is carried out for a new set of lines the circle is of the same diameter and touches the reciprocal lattice at the same origin O but its plane is rotated. The full construction thus consists of a sphere, as it were sitting on the plane of the reciprocal lattice, the point of contact being the origin O. This sphere is pierced by lattice rods arising from rows of atoms sufficiently widely spaced to allow the rods to be of separation less than **k**. Vectors drawn from the real lattice to the points of intersection of lattice rods and the Ewald sphere give the directions in real space in which diffracted beams of electrons can be observed.

Let us now take an example. Consider a rectangular lattice in real space with lattice vectors **a** and **b**. Its reciprocal lattice is also rectangular, with vectors **a*** and **b***. Both of these are depicted in Fig. 4.16 as perpendicular to the plane of the paper. Now the rows along the **a** direction are indexed as (0, 1) since the *x* axis is taken as perpendicular to the plane of the paper (see Fig. 4.4). For clarity a single lattice rod arising from these rows is illustrated and the associated diffracted beam is labelled as the (0, 1). By symmetry there will be a second diffracted beam (as illustrated in Fig. 4.15 at an equal and opposite angle to the incident beam, which is appropriately labelled (0, −1). Lattice rods arising from rows in the **b** direction, which is indexed (1, 0), are in a plane perpendicular to the (0, 1) and (0, −1) rods and thus lie above and below the plane of the paper. Vectors drawn from the sample to the cutting points of these lattice rods with the Ewald sphere give beams, which would be correspondingly labelled (1, 0) and (−1, 0). As well as being diffracted, the electrons may simply be backscattered along their original direction. The

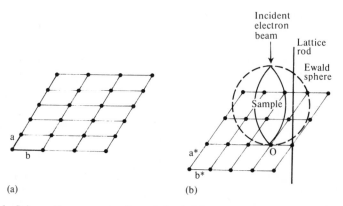

Fig. 4.16. Schematic representation of the full Ewald construction (for clarity one lattice rod only is shown): (a) rectangular real lattice points in a plane perpendicular to the paper; (b) reciprocal lattice points in a plane perpendicular to the paper.

resulting reflected beam is the $(0, 0)$. It is masked partially or wholly by the incoming beam, unless the crystal is tilted by a few degrees.

Other diffracted beams may also be observable if experimental conditions allow. For example, the next group of rows which need to be considered are the diagonals to the rectangular surface mesh. These can be labelled $(1, 1)$ and $(1, -1)$. As before, each set of rows gives rise to two diffraction beams, which are labelled $(1, 1)$, $(-1, -1)$ and $(1, -1)$, $(-1, 1)$. We shall consider the experimental consequences of these diffracted beams when discussing the apparatus for LEED in the next chapter.

Thus far we have discussed the properties of metallic surfaces in geometric terms since the positions of diffracted beams can be interpreted quite readily in this way. Indeed the key feature of the geometric approach is that it brings out the important result that the positions of diffracted beams depend upon the reciprocal lattice rather than the direct lattice. However, an understanding of the diffraction process and particularly the factors which influence the intensities of diffracted beams also calls for some appreciation of the behaviour of electrons in metals and of the nature of the interactions between these electrons and those of the beam. We shall turn our attention to these matters next.

The behaviour of electrons in metals

In the discussion of metallic structures the model taken was that of a close-packed array of spheres held together by non-directional forces. Our purpose now is to elucidate some of the properties of the itinerant electrons that provide the cohesive forces.

1. The free-electron model

As an initial approximation a bonding electron in a metal can be considered as an essentially free particle, able to travel freely through the lattice. In this model the potential energy of the electron is constant and can be conveniently set equal to zero. If now we write the Schrödinger equation

$$H\psi = E\psi$$

in the form appropriate to the motion in just one direction in the lattice, we have

$$H = -\frac{\hbar^2}{2m} \frac{\mathrm{d}^2}{\mathrm{d}x^2}$$

where $\hbar = h/2\pi$ and

$$\frac{\hbar^2}{2m}\frac{d^2\psi}{dx^2} + E\psi = 0$$

where E is now the kinetic energy of the electron.

A solution to this differential equation is

$$\psi = \psi_0 \exp(ikx)$$

where $i = \sqrt{(-1)}$ and $k^2 = 2mE/\hbar^2$. This result can be verified by double differentiation of ψ. As applied to electrons in a metal, this equation describes a one-dimensional free electron with momentum $p = \hbar k$, where k is a wavenumber. In three dimensions, the generalization is

$$\psi = \psi_0 \exp(i\mathbf{k} \cdot \mathbf{r})$$

with $\mathbf{k}$ now the wavevector having Cartesian components (k_x, k_y, k_z) which can be used to label the electron energies as described below. Returning to the one-dimensional case, suppose first that the electrons are treated as being confined to a one-dimensional box of side L. The constraint has now been applied that $\psi = 0$ outside the box and therefore also at the edges of the box. It follows then that $\psi = 0$ at $x = 0$ and $x = L$. Note that if $\psi \neq 0$ at the edges of the box the particle can 'escape' and the boundary condition fails.

Now

$$\exp(i\theta) = \cos\theta + i\sin\theta. \tag{4.1}$$

However, $\cos(0) = 1$, so the cosine term is not acceptable since $\psi(x = 0) = 0$ for this term. Thus the acceptable solution is

$$\psi = \psi_0 \sin kx.$$

However, the requirement that $\psi = 0$ when $x = L$ limits the values of k to the condition that

$$k = \frac{n\pi}{L} \quad (\sin n\pi = 0)$$

where n is a positive integer. Although $n = 0$ satisfies the boundary conditions it also makes $\psi = 0$ and has no physical significance. The allowed values of the electron energy now become

$$E = \frac{\hbar^2}{2m}\left(\frac{n\pi}{L}\right)^2$$

where n is a positive integer.

For an electron moving in a three-dimensional cubic box, the single quantum number n is replaced by $n_x^2 + n_y^2 + n_z^2$. These three quantum numbers, together with spin $s = \pm\frac{1}{2}$, are analogous to the four atomic quantum numbers l, n, m, and s.

An alternative way of applying physically realistic constraints to the freedom of the electrons is to recognize that the perodicity of the lattice implies that if we consider some arbitrary distance a (reverting to the one-dimensional model) over which the lattice repeats, then it follows that ψ also repeats over this length,

$$\psi(x) = \psi(x + a). \tag{4.2}$$

Another way of looking at this is to imagine taking a line of length a and bending it to form a circle of circumference a. Then ψ must join up round the ring, in accordance with eqn (4.2).

We now require that the solution

$$\psi = \psi_0 \exp(ikx)$$

should have the property that

$$\psi(x) = \psi(x + a).$$

Bearing in mind the trigonometrical form of $\exp(ikx)$ (eqn (4.1)) and considering, for example, the cosine term, the condition that

$$\cos kx = \cos k(x + a)$$

is fulfilled provided that $ka = \pm 2n\pi$ where n is a positive or a negative integer. The two values of k correspond to electrons going in opposite directions, although with the same energy given by

$$E = \frac{\hbar^2}{2m} \left(\frac{2n\pi}{a} \right)^2.$$

In three dimensions, the quantum number n can be regarded as constituted from the three quantum numbers n_x, n_y, and n_z relating to the three directions.

It should be noted that the side L of the box or the repeat distance a can be of macroscopic rather than atomic dimensions. The result in either treatment is that the allowed values of E are close together. Indeed the energy levels essentially form a continuum.

There is one other property of the electronic states of a metal which is of some importance, namely the number of states with any particular energy. This is known as the density-of-states function $N(E)$, which for free electrons can be shown to be given by

$$N(E) \propto E^{1/2}.$$

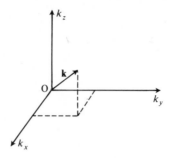

Fig. 4.17. **k**-space diagram: electrons with low values of **k** are near the origin and those with the highest energy form the Fermi surface.

If we bear in mind that electrons obey Fermi–Dirac statistics, so that no two electrons are allowed to have all four quantum numbers (n_x, n_y, n_z and s) the same, the energy levels in a metal can be filled up according to the *aufbau* principle, just as in atoms. Each level, defined by a particular value of n, thus accommodates two electrons. A convenient way of describing this process is to consider a three-dimensional space whose mutually perpendicular axes represent the components of the wavevector **k**. The three axes are thus k_x, k_y, and k_z, as shown in Fig. 4.17. Any point in the resulting **k** space defines the crystal momentum. Now as we fill in the electrons, those with low quantum numbers (and thus little momentum) have low energy and are located close to the origin, whilst those with high energy are far from the origin. Thus in **k** space filling in the electrons is like pumping up a balloon whose centre is the origin. The electrons with the highest energy are those at the surface, which is known as the Fermi surface. Their energy is known correspondingly as the Fermi energy E_F. E_F is typically 1–10 eV in magnitude.

The *aufbau* process can also be represented using the density of states function $N(E)$. Electrons occupy the $N/2$ lowest states (N being the number of electrons) and the higher energy states are empty. We may note that if the metal is heated some thermal excitation to empty states adjacent to the Fermi surface occurs. However, since E_F is large compared with the temperatures normally encountered ($1\ \text{eV} \equiv 1.2 \times 10^4\ \text{K}$), the fraction excited is very small. The electron distribution is illustrated in Fig. 4.18.

2. Band model

Thus far the influence of the lattice on the predicted behaviour of the electrons has been minimal. However, we have to recognize that the

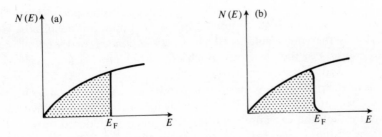

Fig. 4.18. Density of states in the free-electron model at (a) 0 K and (b) elevated temperatures. The shaded areas are occupied states and E_F is the Fermi energy.

potential experienced by the electrons must be expected to change, perhaps sharply, near the atom cores. If this happens the potential energy does not have the constant value assumed in the free-electron model. An alternative way of describing this situation is the model of Kronig and Penney, which is illustrated in Fig. 4.19 for a one-dimensional lattice. The potential wells due to the atoms introduce a periodicity, characteristic of the atomic array, into the wavefunctions for the electrons since the probability density $\psi\psi^*$ is periodic with the period of the lattice. A theorem due to Bloch allows the total wavefunction to be expressed as a combination of a periodic function characteristic of the lattice and a plane wave such as that encountered in the free-electron model. The resulting wavefunction is called a modulated plane wave and is expressed by

$$\psi_{\text{total}} = U_{\mathbf{k}}(\mathbf{r})\exp(i\mathbf{k} \cdot \mathbf{r}) \tag{4.3}$$

where $U_{\mathbf{k}}(\mathbf{r})$ is the periodic function. This total wavefunction fulfils the periodicity requirement and simplifies correctly to $\exp(i\mathbf{k} \cdot \mathbf{r})$ when the

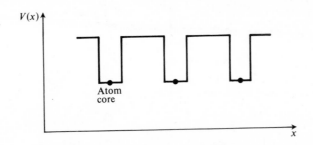

Fig. 4.19. One-dimensional Kronig–Penney model showing the variation of the potential $V(x)$ with distance x in a metal crystal.

lattice potential $U_k(\mathbf{r})$ is 'switched off', i.e. when it becomes constant. The treatment of eqn (4.3) is complex; the most important result for our discussion is that the resulting Bloch wavefunctions show that there are certain energies of the electrons which are forbidden. These energies are known as band gaps.

A more pictorial representation of band gaps can be given using the **k**-space diagram of Fig. 4.17. If we imagine an electron with wavevector **k** moving through the crystal, its motion is in general hindered only by the vibrations of the atomic cores. However, in some directions the electron may encounter a set of planes which fulfil the Bragg condition

$$n\lambda = 2d \sin \theta.$$

When this happens, the electron is reflected and therefore no longer propagates through the lattice. The result is to introduce a discontinuity into the plot of E versus k shown in Fig. 4.20. This plot is known as a dispersion curve and is parabolic in shape because, as we have already seen,

$$k^2 = 2mE/\hbar^2.$$

Detailed analysis shows that the values of **k** for which the electron cannot propagate lie on a series of planes which join up to form a box, the shape of which is determined by the geometry of the lattice. This box is known as a Brillouin zone. It then follows that electrons can only exist in states which lie within a Brillouin zone. We can now imagine once again carrying out the *aufbau* process, but this time taking account of effect of Brillouin zones on the Fermi surface. At low **k** values the Fermi surface is essentially spherical, just as in the free-electron model. However, when the addition of electrons causes the Fermi surface to touch the planes of

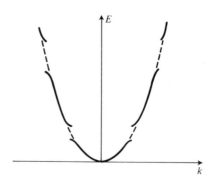

Fig. 4.20. Dispersion curve showing band gaps: - - - -, free-electron model.

the Brillouin zone, i.e. the walls of the box, it is no longer completely free to expand. Additional electrons are limited to filling in the corners of the box. As the result fewer electrons can be accommodated for each increment in E. The plot of $N(E)$ against E turns down when this happens. Eventually the zone is full. Any further electrons must go into the next Brillouin zone, which is separated from the first by an energy gap. The filling of the first zone can be likened to pumping up a balloon in a rigid box. At first the balloon expands maintaining its normal spherical shape. When it touches the sides of the box, further pumping causes the balloon to adapt to the shape of the box. Eventually, when the balloon fills the box no more inflation is possible. This process can be looked at in terms of the band model, when the effects of Brillouin zones on the density-of-states function $N(E)$ are as illustrated in Fig. 4.21.

So far we have not given any consideration to the way in which the bands are formed from the wavefunctions of the individual atoms which assemble to form the metal. The approach which considers the build-up of bands from atomic orbitals is called the 'tight-binding approximation'. As a starting point we can consider the interaction between a valence orbital on each of two atoms of the same element. Molecular orbital theory indicates that the result is the formation of two molecular orbitals, one of higher energy than the atomic orbitals (i.e. antibonding) the other of lower energy (i.e. bonding). Bringing up a third and subsequent atoms, each with a bonding orbital, produces molecular orbitals involving all the atoms and with ever decreasing energy spacings. Eventually, the spacings become so fine-grained as to constitute the continuum we have called a band. We should note that each set of valence orbitals on the constituent atoms forms a band when the atoms assemble to form the bulk metal. In practice two important situations may arise. In the first the energy separation between the valence orbitals on the atoms is so great that when the atoms congregate to form the metal the resulting bands are well separated in energy. This is the situation already illustrated in Fig. 4.21.

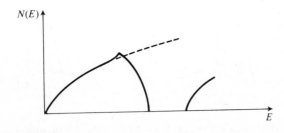

Fig. 4.21. Density of states in the band model: - - - -, free-electron model.

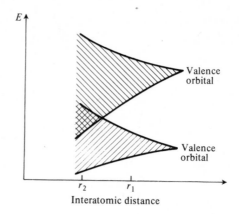

Fig. 4.22. Formation of bands from two atomic valence orbitals. An equilibrium interatomic spacing of r_1 gives two non-overlapping metallic bands; at r_2 overlapping bands are formed.

In the second situation a smaller separation of energies in the atoms may lead to overlap of bands in the metals. The formation of overlapping bands, as illustrated in Fig. 4.22, at the smaller atomic separation, r_2 is characteristic of transition metals. The resulting density-of-states diagrams for two typical examples, nickel and copper, are shown schematically in Fig. 4.23. Note that the strongly interacting 4s electrons form a broad band, whereas the less readily ionized and therefore less strongly interacting 3d electrons form a narrow band. Also note that in copper the 3d band is full, whereas in nickel there is about half an electron per atom missing from this band.

Thus far we have considered the theory of electrons in the bulk of the metal. However, the question arises of whether such a description is appropriate at the surface. We consider this point next.

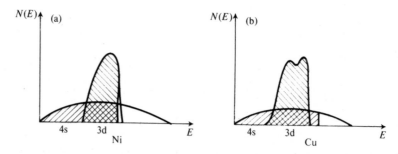

Fig. 4.23. Schematic diagram of densities of states for (a) nickel and (b) copper.

Surface wavefunctions

The wavefunctions used to describe the electrons in a metal, the Bloch waves of eqn (4.3), are not expected to be a correct description at the surface because of the abrupt change in the potential experienced by the electrons as they reach the surface. Basically, the bulk wavefunctions must die away rapidly but smoothly (wavefunctions are always continuous so a stepped wavefunction is not acceptable) into the space beyond the surface. In addition, there are wavefunctions which arise from the boundary conditions caused by the surface. These surface wavefunctions take the two typical forms illustrated in Fig. 4.24. When a 'surface state' is formed (Fig. 4.24(a)) the wavefunction has considerable amplitude at the surface but dies away rapidly into the bulk. A 'surface resonance' (Fig. 4.24(b)) is characterized by a considerable amplitude at the surface and a small but continuing amplitude into the bulk.

These surface wavefunctions may have a significant effect on the density of states at the surface. For example, in the extreme case where the surface state is completely localized on the surface atom and is occupied, the requirement for charge neutrality of the surface atom is satisfied. The electrons in the band characteristic of the bulk then have no density at the surface. In effect electrons have become completely localized at the surface. Surface states will, in general, arise when the surface parameters differ significantly from the bulk. They are known as Tamm states.

A further consequence of the abrupt change in potential when the crystal terminates may be to allow wavefunctions to become acceptable at the surface which are not allowed in the bulk. The result is that there are energy levels attributable to the surface located in the energy gap between two bulk bands. These are known as Shockley states. It is worth

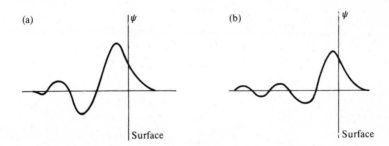

Fig. 4.24. Schematic representation of surface wavefunctions: (a) surface state; (b) surface resonance.

noting that the unambiguous identification of a state as being due specifically to the surface is by no means straightforward. Since such a state is identified as resulting from the difference between some property, usually associated with the emission of electrons, observed experimentally and the predictions of theory, it is only when a detailed concordance between theory and experiment relating to the bulk has been achieved that the assignment of surface states can be made. Furthermore, each face of each metal has its own electron emission characteristics, so that no generalizations are expected. Rather, each face has to be the subject of an individual calculation and comparison with experiment.

We now have an outline of the properties of the electrons in a metal and can consider the possible outcomes of bombardment of the surface by a beam of low energy electrons. Our main concern will be with diffraction, but since this represents only a small fraction of the interactions, the other possible fates of impinging electrons will first be considered briefly.

5. Low energy electron diffraction

Introduction

The interaction between fundamental particles subject to the wave–particle duality, such as photons or electrons, and a metal is a complex process. Descriptions of the results are to be found expressed in purely geometric terms, in classical wave theory, in the language of wave mechanics, and in terms of scattering theory. Each of these approaches has advantages for the best understanding of different aspects of LEED and each will be used as is most appropriate. We may note first that in the discussion of the Ewald construction in the previous chapter (a geometric approach), diffracted beams were identified at the points of intersection of the lattice rods and a sphere of radius equal to the magnitude of the wave vector $\mathbf{k} = 1/\lambda$. This implies that the wavelength of the electrons does not change, i.e. diffraction is a property of elastically scattered electrons. However, other, and inelastic, processes are much more probable than elastic scattering and some preliminary account of these will put LEED in context.

Inelastic electron scattering

As the electron approaches the metal surface it first encounters the valence band electrons, which it may excite. The excitation may be sufficient to cause ejection of valence electrons from the surface (secondary emission). Although it is possible to collect these electrons, the voltages in a LEED apparatus are such as to turn them back to the specimen. The net result is an inelastic scattering process. Another and important possibility for excitation and inelastic scattering is that the incoming electron raises the metal electrons to excited states by the production of surface or bulk plasmons, collective electron oscillations within the sample, or interband transitions.

Further penetration of the electron brings it to the atom cores. A possible low energy interaction with these is the excitation of vibrational modes of the lattice. However, the quanta involved are much smaller than the energy of the electrons and reflected electrons which have undergone such phonon energy exchange cannot be distinguished from truly elastically scattered electrons. This is therefore called 'quasi-elastic' scattering.

At the other extreme of energy, a highly energetic electron may cause the ejection of a core electron from the lattice atom. The resulting excited ion may relax in a variety of ways, one of which is by the ejection of a secondary electron. This process, which is of low probability, is known as the Auger effect and will be discussed in Chapter 6. From the point of view of LEED, the significant interaction between the electron and the atom cores is one which completely reverses the direction of the electron, leading to elastic back scattering and, therefore, to the possibility of diffraction. However, only a small fraction of the electrons are thus back scattered, the majority being forward scattered and lost, eventually, to the sample. Typically, no more than 2 per cent of the electrons in the original beam appear in a diffracted beam.

Some idea of how many atom layers are involved in these scattering processes may be given by noting that the cross-section σ for both plasmon scattering and ion-core scattering is $\sim 10^{-20}\,\mathrm{m}^2$ (though this figure depends on the electron energy). The mean free path l of an electron is given by

$$l = (\rho\sigma)^{-1}.$$

where ρ is the atom density, which is typically 10^{29} atom m^{-3}. Therefore

$$l \sim 10^{-9}\,\mathrm{m}.$$

Thus scattering is a property of the top few layers. This conclusion is reinforced by consideration of the photoemission of electrons discussed in Chapter 6. As Fig. 6.13 shows, LEED uses electrons at the minimum of the curve where the mean free path is shortest. In order to describe further the ion-core scattering of electrons, the process which reverses the direction of the electrons and is therefore fundamental to diffraction, a model of the electric potential within the crystal is required. This is discussed next.

The crystal potential

We consider now the electrical interactions as an incident electron penetrates the lattice, that is to say the variation of the potential within the crystal. This can be divided into two components.

1. A 'general' potential which does not take specific account of the atomic structure of the lattice and leads to the series of energies in **k** space already discussed. This potential stops the metallic electrons from escaping and may be related to the band model as shown in Fig. 5.1. As can be seen

$$V_0 = \phi + E_\mathrm{F}$$

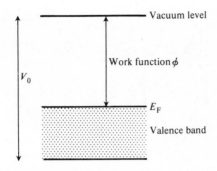

Fig. 5.1. The 'general' potential in a metal.

The potential V_0 is known as the 'inner potential'. Typically $E_F \approx 10$ eV and $\phi \approx 5$ eV so that $V_0 \approx 15$ eV. The work function ϕ is the energy required to remove an electron from the Fermi level into a vacuum and can be looked on as the solid state analogue of the ionization potential of an atom.

2. An ion-core potential arising from the atomic structure of the lattice. This potential is due to the electrons remaining on the atom and to the nuclear charge. It is spherically symmetrical. The interaction of the ion core with the electron is assumed to be coulombic and leads to the elastic scattering required for diffraction. At this stage two problems are encountered: (i) how to relate the potentials of adjacent ion cores and (ii) how to join up the general and the ion-core potentials. The solution to these problems is achieved by using an approximate but mathematically tractable model in which the largest possible non-overlapping spheres are drawn round each atom. The detailed form of the potential within the atoms is not required, though it approximates to the coulombic form $V(r) = -Z/r$. What is required is the scattering power of the atom; this is parametrized for computational purposes in the 'phase shift' discussed later. Outside the spheres the rest of the crystal has the constant inner potential V_0. Thus if a section is taken through the crystal in the z direction within the bulk, the potential alternates between the ion core and the general potential. There is a step at the surface. The result is illustrated in Fig. 5.2. For reasons which are obscure to those unfamiliar with North American kitchen equipment, the model has been christened the 'muffin-tin' model.

This model provides a successful basis for the detailed treatment of diffraction in terms of scattering theory, and is particularly useful for calculation of the intensities of diffraction spots. A definitive account is to be found in Pendry (1974).

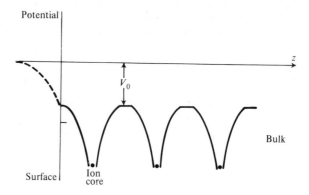

Fig. 5.2. Variation of the potential in a metal.

After this consideration of the varied interactions between an electron beam and a metal we turn to the experimental method and some results of LEED.

The LEED experiment

In planning a LEED experiment the first matter for decision is the energy for the monochromatic electron beam. Two considerations immediately present themselves. The first is that the de Broglie wavelength λ of the electrons should be appropriate. This property is given by

$$\lambda = h/p$$

which becomes

$$\lambda = \left(\frac{150}{V}\right)^{1/2} \text{Å}$$

$(1\,\text{Å} = 10^{-8}\,\text{cm})$ or

$$\lambda = \left(\frac{1.5}{V}\right)^{1/2} \text{nm}$$

where V is the voltage through which the electrons have been accelerated.

The second requirement, which arises from an intention to use LEED as a surface-sensitive technique, is that the electrons shall not penetrate too deeply into the specimen. Electrons with energy in the range 20–500 eV fulfil these two criteria.

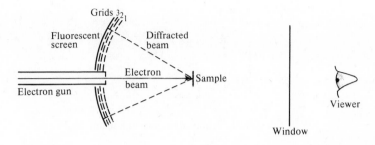

Fig. 5.3. Schematic representation of a LEED apparatus. The applied potentials are as follows: grid 1, the final electrode of the electron gun and the sample are at earth potential so that the electrons move in a field-free region; grid 2 is slightly negative relative to the primary electron beam and therefore turns back inelastically scattered electrons; grid 3 is at several thousand volts positive potential and accelerates the electrons so that they activate the fluorescent screen.

The apparatus consists of a vacuum chamber, usually constructed of stainless steel, which can be evacuated to the ultrahigh vacuum region. A diagrammatic representation is shown in Fig. 5.3. The sample is cut from a single crystal ingot so that a crystal plane of the required orientation is exposed. It is mounted at the centre of the vacuum chamber. A typical sample would be a thin wafer of dimensions roughly $1\,\text{cm} \times 1\,\text{cm} \times 0.01\,\text{cm}$. The electron beam, which is designed to be rather accurately parallel, usually impinges normally on the surface and a system of gauze grids ensures that only elastically back scattered, i.e. diffracted, electrons are recorded at the detector. This is a fluorescent screen onto which the electrons are accelerated by a potential of several thousand volts. With this energy the electrons produce a bright spot which can be seen and photographed. For intensity measurements, the brightness of the spots can be measured with a photometer or the electrons can be collected and their current recorded as a function of voltage.

Plane-wave theory of LEED

In addition to the approaches discussed so far (via the Ewald construction or the muffin-tin model) the diffraction process can also be considered from a wave-like point of view. In this approach the beam of electrons is treated as a plane wave incident normally on the surface. Diffraction then occurs when there is constructive interference between the electrons back scattered from lattice points. If we consider first diffraction by a row of

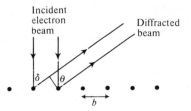

Fig. 5.4. Diffraction in one dimension.

lattice points of separation b, as illustrated in Fig. 5.4, then for reinforcement

$$\delta = k\lambda = b \sin \theta_k \quad (k = 0 \text{ or an integer}). \tag{5.1}$$

There is no limitation on the orientation of the diffracted beams; they are subject only to the condition that θ_k is constant. The diffraction pattern from a set of points constituting a one-dimensional lattice is therefore a series of cones whose angles are defined by eqn (5.1), (Fig. 5.5). If we now consider the diffraction condition for a second set lattice points of separation a, then a condition analogous to eqn (5.1) will apply, namely

$$h\lambda = a \sin \theta_h \quad (h = 0 \text{ or an integer}). \tag{5.2}$$

This equation describes a second set of cones. If now the two sets of lattice points are chosen so that together they constitute the surface, a

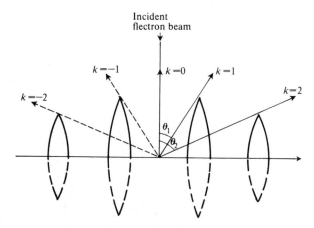

Fig. 5.5. Diffraction cones in one dimension fulfilling the condition $k\lambda = b \sin \theta$ (eqn (5.1)). In practice electrons will not be observed below the plane of the crystal as indicated by the broken lines.

LEED experiment will record beams in directions which satisfy eqns (5.1) and (5.2) simultaneously, i.e. where the two sets of cones intersect.

In the discussion of diffraction and of the Ewald construction in the previous chapter, we noted the significance of the reciprocal lattice for interpreting diffraction patterns. It was asserted, without proof, that diffracted beams would occur in directions determined by the points at which the lattice rods intersected the Ewald sphere. Let us now compare the reciprocal lattice approach with the wave treatment just given and show that the assertion was justified.

Comparison of reciprocal lattice and plane-wave interpretations of LEED

In bringing together the two treatments of diffraction, we note first that when carrying out the Ewald construction as described in the previous chapter, the Ewald sphere was defined as being centred on the crystal and as having a radius of $1/\lambda$. Now in a LEED apparatus, the fluorescent

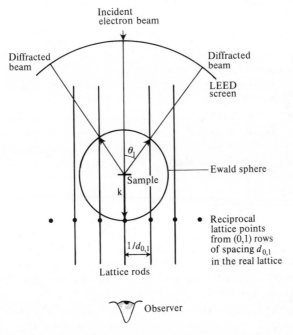

Fig. 5.6. Schematic representation of the Ewald construction and the plane-wave interpretation of diffraction.

detecting screen forms part of a spherical surface with the sample at the centre of the sphere. Thus the Ewald sphere and the screen have a common centre, at which the sample is located.

For the purposes of illustration, the situation can be depicted as in Fig. 5.6, taking again a single row of reciprocal lattice points for clarity. The lattice rods in reciprocal space have a spacing $1/d$ corresponding to lattice point spacings of d on the surface.

Then for the first diffracted beam eqn (5.2) can be written as

$$\lambda = d \sin \theta_1.$$

Now turning to the reciprocal lattice, from Fig. 5.6 we can see that

$$\sin \theta_1 = \frac{1/d}{1/\lambda} \quad (\mathbf{k} = 1/\lambda) \tag{5.3}$$

or, once again,

$$\lambda = d \sin \theta_1.$$

If the incoming beam had been of shorter wavelength, so that $1/\lambda$ was larger, a second lattice rod would have been cut, and in addition to eqn (5.3) we should have had the possibility

$$\sin \theta = \frac{2/d}{1/\lambda}$$

which corresponds to $h = 2$ in eqn (5.2).

There is, therefore, satisfactory agreement between the two approaches. We shall return to the Ewald model in order to consider in more detail what form the pattern of spots which go to make up a LEED pattern is expected to take. This approach serves to emphasize again that the important relationship is between the LEED pattern and the surface reciprocal lattice (rather than the direct lattice).

Indexing LEED patterns

Let us consider first an Ewald circle such as that shown in Fig. 5.6. The lattice rod spacing $1/d_{01}$ is determined by the rows of atoms of spacing d_{01} with index (01). These rows are perpendicular to the plane of the paper. In conformity with the plane-wave theory, we note that the index (01) means that $h = 0$ and $k = 1$. Then $\sin \theta_h = 0$ and $\theta_h = 0°$; that is to say the beam angle is unchanged in the x direction, i.e. it is perpendicular to the plane of the paper. Now the lattice rods cut the Ewald circle at two positions whose angles are described by θ and $-\theta$. Accordingly the spots are labelled $(0, 1)$ and $(0, -1)$ respectively. From the point of view of an observer of the screen, who is situated as shown, two bright spots will be seen which are symmetrically placed with respect to the reflected beam, which is indexed as $(0, 0)$. Furthermore, the smaller the value of $1/d_{01}$ the closer are the spots to the incident beam. This leads to a conclusion, important when adsorbed layers are studied, that widely spaced lattice points in the real lattice give closely spaced diffraction spots (high $d_{01} \rightarrow$ low $1/d_{01} \rightarrow$ close spots).

We now consider reciprocal lattice rods along a line perpendicular to that just discussed, i.e. perpendicular to the plane of the paper. These rods are derived from the (10) rows, and again cut the Ewald sphere in two places. Therefore two more spots, symmetrically disposed with respect to the $(0, 0)$ beam but lying above and below the plane of the paper, are anticipated. These are the $(1, 0)$ and $(-1, 0)$ spots. Their spacings are proportional to $1/d_{10}$. The diagonal row (11) gives diagonally disposed spots $(1, 1)$ and $(-1, -1)$ and the other diagonal row (-11) gives spots $(-1, 1)$ and $(1, -1)$. In practice these are the spots most commonly encountered and an indexed pattern is illustrated in Fig. 5.7. Because the spot indices involve only whole numbers, the spots are called 'integral order'.

The positions of diffraction spots from clean metals are of some intrinsic interest, particularly on the question of whether the lattice spacings on the surface, which in this case are atomic distances, are the same as in the bulk as separately determined by X-ray diffraction. In most cases there is less than 10 per cent difference. However, of more general concern is the development of new patterns when adsorption occurs. A system of nomenclature is required for these overlayers and this is described next.

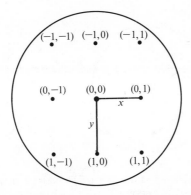

Fig. 5.7. Indexing diffraction spots: the pattern from a rectangular lattice. The distance x is proportional to $1/d_{01}$; y is proportional to $1/d_{10}$.

Nomenclature of overlayer lattices

The method of nomenclature can be most readily followed if we take an example and consider the change in the LEED pattern which might be brought about by adsorption. Suppose we start with an oblique surface lattice, with surface vectors **a** and **b**, as illustrated in Fig. 5.8(a). The corresponding reciprocal lattice is in Fig. 5.8(b) (note that $\mathbf{a} \perp \mathbf{b}^*$ and $\mathbf{b} \perp \mathbf{a}^*$). As we have seen, the diffraction pattern has the same shape as the reciprocal lattice. The number and spacing of the diffraction spots observed depends upon the radius of the Ewald sphere, i.e. the wavelength of the electrons. Trivially, the spots observed could also depend upon the size of the fluorescent screen; it is assumed to be adequate. Let us suppose that the pattern is as shown in Fig. 5.8(c). Following adsorption the pattern changes to that of Fig. 5.8(d) where just one rhombus is shown for clarity. For the overlayer lattice we can write

$$\mathbf{b}^{*\prime} = \mathbf{b}^*$$
$$\mathbf{a}^{*\prime} = \mathbf{a}^*/2,$$

which implies that in the real lattice

$$\mathbf{b}' = \mathbf{b}$$
$$\mathbf{a}' = 2\mathbf{a}.$$

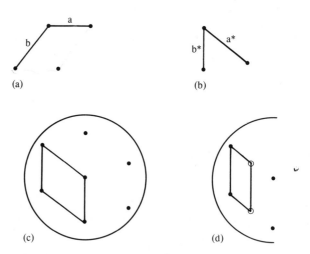

(a)   (b)

(c)   (d)

Fig. 5.8. Schematic representation of LEED pattern changes due to overlayer formation (●, original LEED spot; ○, extra LEED spot ($\frac{1}{2}$, 1): (a) real lattice; (b) reciprocal lattice; (c) LEED pattern from a clean surface; (d) LEED overlayer pattern (part).

Therefore the overlayer unit cell is twice as long in one direction as the metal unit cell. However, it should be noted that it is not possible to tell from this pattern how the overlayer is related geometrically to the underlying atoms. Furthermore, it should not be thought that the overlayer pattern is composed of two parts, the original spots due to the underlying metal atoms plus spots from the overlayer. The new pattern is characteristic of the new surface. However, since the positions of the extra spots can be described by fractions of the original spacings ($\mathbf{a}^{*\prime} = \mathbf{a}^*/2$), they are known as 'fractional-order' spots.

The nomenclature commonly used to identify the overlayer pattern is that due to Wood. In a case where the unit vectors of the overlayer are parallel to the vectors of the substrate, they are described by

$$\mathbf{a}' = n\mathbf{a} \quad \text{and} \quad \mathbf{b}' = m\mathbf{b}$$

where n and m are integers; in our example $n = 2$ and $m = 1$. The overlayer is then said to be primitive and to have a $p(n \times m)$, i.e. $p(2 \times 1)$, mesh. The overlayer can be more fully described by defining the metal M, the particular plane under consideration (h, k, l), the adsorbate X, and the fractional coverage θ, i.e. in full

$$M(h, k, l)(n \times m)\text{--}X\text{--}\theta.$$

Overlayer mesh vectors are not always parallel to the substrate, but are quite often rotated through an angle. When this happens the angle of rotation is also quoted. One example of this which is frequently encountered is rotation through $45°$. A particular case which illustrates the way LEED patterns develop is the adsorption of oxygen on the tungsten (100) plane, which will now be discussed.

We begin by considering the pattern from clean tungsten (100). This has a square lattice (i.e. $\mathbf{a} = \mathbf{b}$) as shown in Fig. 5.9(a) and so gives rise to the pattern shown in Fig. 5.9(b). When oxygen is adsorbed, the first new pattern is as illustrated in Fig. 5.9(c). From this figure it can be seen that the new reciprocal lattice is given by

$$\mathbf{a}^{*\prime} = \mathbf{a}^*/2 \qquad \mathbf{b}^{*\prime} = \mathbf{b}^*/2.$$

Thus in the direct lattice $\mathbf{a}' = 2\mathbf{a}$ and $\mathbf{b}' = 2\mathbf{b}^*$, so that this is a $p(2 \times 2)$ structure. Further exposure gives rise to the pattern illustrated in Fig. 5.9(d). The primitive reciprocal lattice cell is indicated and Pythagorean geometry gives

$$\mathbf{a}^{*\prime} = \mathbf{a}^*/\sqrt{2} \quad \text{and} \quad \mathbf{b}^{*\prime} = \mathbf{b}/\sqrt{2}.$$

In the direct lattice the primitive unit cell is therefore of side $\sqrt{2}\mathbf{a}$ and

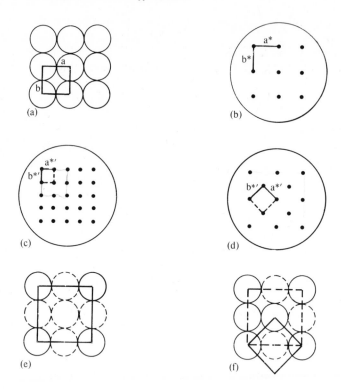

Fig. 5.9. Surface structures and LEED patterns for the tungsten (100)-oxygen system: (a) clean tungsten (100) surface and unit cell; (b) LEED pattern of clean surface; (c) LEED pattern with spots at half the spacing of (b); (d) LEED pattern with spots at $\sqrt{2}/2$ times the spacing of (b); (e) possible surface structure and unit cell corresponding to (c); (f) possible surface structure corresponding to (d). The unit cell can be drawn in either of two ways as shown: ——, $(\sqrt{2}\times\sqrt{2})R45°$; – – –, $c(2\times2)$.

$\sqrt{2}\mathbf{b}$ and rotated through 45°. It could therefore be designated as $(\sqrt{2}\times\sqrt{2})R45°$. However, if the primitive unit cell is extended as shown in Fig. 5.9(e) the adsorbed layer has a structure similar to the $p(2\times2)$ with the addition of an extra lattice point at the centre of the square. This pattern is of common occurrence and it is therefore given its own designation as centred (2×2), written $c(2\times2)$. In this example the vectors $\mathbf{a}^*$ and $\mathbf{b}^*$ are equal throughout, as are $\mathbf{a}^{*\prime}$ and $\mathbf{b}^{*\prime}$, although this is not a necessary condition for the formation of the $c(2\times2)$ structure.

Again it should be noted that the LEED pattern gives the geometry of the unit cell following adsorption, but not the registry between this layer and the underlying metal atoms. Experience has shown that many gases

 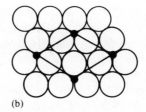

Fig. 5.10. Overlayer structures: (a) p(2×1); (b) (√3×√3)R30°. It should be noted that the LEED patterns do not define the registry between the metal atoms at the surface and the adsorbed species.

give layers whose surface vectors are in simple numerical ratios to the clean surface meshes. Among the most common, in addition to the two discussed above, are (1×1), (2×1) and (√3×√3)R30°, of which the latter two are illustrated for some low index planes in Fig. 5.10.

Occasionally the integers n and m may be much larger than the values so far mentioned. For example, ammonia adsorbed on tungsten (211) gives a layer with structure (7×2). It is not easy to envisage a mechanism by which order at such long range could be achieved and an alternative explanation is sought. It is suggested that the adsorbed layer has a symmetry of its own, which is unrelated to the symmetry of the substrate. However, every so often the scattering points of the overlayer will come into coincidence with the surface atoms. This will happen each time an integral number of unit mesh vectors of the substrate, say $Y\mathbf{a}$, equals an integral number of unit meshes of the overlayer, say $Z\mathbf{a}'$, i.e.

$$Y\mathbf{a} = Z\mathbf{a}'.$$

When $Y = 1$ we have a normal layer, otherwise the layer is described as a 'coincidence lattice'. The justification for accepting that such a coincidence lattice will indeed give a pattern with high Wood indices has to have recourse to consideration of the factors which influence the intensity of diffraction spots (see Van Hove and Tong 1979).

The coherence length and surface reconstruction

In our discussion of diffraction so far, we have made two implicit assumptions. The first is that the single-crystal surface is perfectly ordered over an indefinitely large area. The second is that the atomic array of a plane within the crystal is not significantly disturbed when that plane becomes the surface. These assumptions are not always justified. As far as surface order is concerned, the relevant question is how large an area of the surface is required to give a satisfactory LEED pattern. The answer depends in part upon the properties of the electron beam. So far, the beam has been treated as though it were a simple plane wave whose properties are uniform across the wavefront. A real electron beam is of course not like this, but is made up of electrons with slightly differing velocities and directions. There is then a characteristic short distance over which the electron beam can be regarded as behaving ideally. This distance is called the 'coherence length' and is roughly 100–500 Å. For a LEED pattern to develop the surface structure must be uniform on this scale. If the periodicity of the surface structure is large compared with the coherence length, the structure does not give a LEED pattern.

As far as our second assumption is concerned, we should not expect it to be exactly correct; some surface relaxation is to be expected to minimize the surface free energy. In most metals the atomic distances in the surface layer change by less than 10 per cent. However, in just a few cases the LEED patterns from clean metals correspond to a unit cell significantly larger than that expected. The surface is then said to be 'reconstructed'. Thus, for example, the platinum (100) surface is metal-stable in the (1×1) structure and reconstructs to a stable (5×20) arrangement.

Intensities of LEED spots

Intensities of diffracted beams are usually measured by including in the LEED apparatus a facility for recording the current carried by each beam. The device for doing this is called a Faraday cup. Two related aspects of beam intensities are of interest. The first is the way in which the intensity varies from beam to beam, and the second is the way in which the intensity of any particular beam varies with the electron energy. We have already seen that the positions of the diffracted beams give information only about the surface geometry. Intensity analysis is required for determining the positions of adsorbed species relative to the substrate. The theories of LEED intensities are of considerable complexity and only an outline will be given. A full account is given by Pendry (1974).

1. Kinematic theory

In this theory it is assumed that the interaction between the ion cores and the electrons, which we have already discussed, is such that any electron only suffers a single scattering event when interacting with the cores. For this reason it is also known as a 'single-scattering' theory. In general, the postulate of single scattering is likely only to be a rough approximation and a more sophisticated theory should take account of the possibility of multiple scattering. On the other hand, the simpler theory gives at least a qualitative insight. Much of the kinematic theory is based on the theory of optical diffraction, which will be summarized next.

2. One-dimensional grating

The theory starts with a one-dimensional treatment in which there are incident and scattered waves defined by the unit vectors $\mathbf{S}_0$ and $\mathbf{S}$. These vectors identify the directions of the waves. The wave with unit vector $\mathbf{S}_0$ is incident at some angle θ_i on a grating consisting of a row of point scatterers with separation defined by a vector $\mathbf{d}$. The outgoing wavevector is at some other angle θ_r (both angles are to the normal). Thus we have the vector diagram shown in Fig. 5.11.

Then the path difference Δ between waves scattered by adjacent atoms satisfies the two conditions

$$\Delta_1 = \mathbf{d} \sin \theta_i$$
$$\Delta_2 = \mathbf{d} \sin \theta_r$$

where

$$\Delta = \Delta_2 - \Delta_1.$$

By the rules of vector multiplication, Δ_1 and Δ_2 are also the scalar products

$$\Delta_1 = \mathbf{d} \cdot \mathbf{S}_0 \qquad \Delta_2 = \mathbf{d} \cdot \mathbf{S}.$$

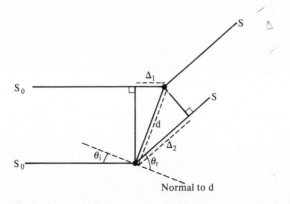

Fig. 5.11. Scattering of a one-dimensional grating of spacing $\mathbf{d}$. $\mathbf{S}_0$ is the incident-wave unit vector and $\mathbf{S}$ is the scattered-wave unit vector.

Therefore the overall path difference is

$$\Delta = \Delta_2 - \Delta_1 = \mathbf{d} \cdot \mathbf{S} - \mathbf{d} \cdot \mathbf{S}_0 = \mathbf{d} \cdot (\mathbf{S} - \mathbf{S}_0).$$

For constructive interference between the outgoing wavevectors, the path difference Δ must be an integral number of wavelengths, that is

$$n\lambda = \mathbf{d} \cdot (\mathbf{S} - \mathbf{S}_0).$$

A 'scattering vector' $\mathbf{K}$ is now defined as

$$\mathbf{K} = \frac{2\pi}{\lambda} (\mathbf{S} - \mathbf{S}_0).$$

Thus

$$2\pi n = \mathbf{d} \cdot \mathbf{K}$$

which is called the Laue condition for constructive interference.

If there are M scattering centres giving rise to interference, the intensity $I(\mathbf{K})$ of the scattered wave is given by

$$I(\mathbf{K}) = \frac{A^2 \sin^2\{(M/2)\mathbf{K} \cdot \mathbf{d}\}}{\sin^2(\tfrac{1}{2}\mathbf{K} \cdot \mathbf{d})} \qquad (5.4)$$

where A is the amplitude of the beam.

3. Two-dimensional grating

We now turn to a two-dimensional array of scattering points. It is necessary to bear in mind at this stage that the unit mesh, that is to say the scattering unit on the surface, may contain more than one atom. If now the vectors defining the unit mesh are $\mathbf{a}$ and $\mathbf{b}$, the Laue interference conditions become

$$\mathbf{K} \cdot \mathbf{a} = 2\pi h \quad (h = 0, 1, 2, \ldots)$$
$$\mathbf{K} \cdot \mathbf{b} = 2\pi k \quad (k = 0, 1, 2, \ldots)$$

and must be satisfied simultaneously. The scattered intensity is then a function of both $\mathbf{K} \cdot \mathbf{a}$ and $\mathbf{K} \cdot \mathbf{b}$. The equation analogous to that for a linear grating (eqn (5.4)) is

$$I(\mathbf{K}) \propto A^2 \frac{\sin^2(\tfrac{1}{2}M_1\mathbf{K} \cdot \mathbf{a})}{\sin^2(\tfrac{1}{2}\mathbf{K} \cdot \mathbf{a})} \frac{\sin^2(\tfrac{1}{2}M_2\mathbf{K} \cdot \mathbf{b})}{\sin^2(\tfrac{1}{2}\mathbf{K} \cdot \mathbf{b})} = \text{I.F.}$$

This expression, which describes the interference phenomenon, is known as the interference function (I.F.). At this stage the kinematic theory predicts that scattering from a two-dimensional array will lead to a pattern of spots of equal intensity. The spacing of these spots depends on the surface geometry, that is to say the surface unit mesh vectors **a** and **b**. However, it is also necessary to take account of the scattering ability of the atoms in the unit mesh which is known as the 'scattering factor'. The amplitude of the scattered wave, when just a single unit mesh is contributing, is called the 'structure factor' F and as defined by

$$F = \sum_{n=0}^{N-1} f_n \exp(-i\mathbf{K} \cdot \mathbf{r}_n).$$

where f_n is the scattering factor for an atom n (and depends on the number of electrons in the atom), N is the number of atoms in the unit mesh and $\mathbf{r}_n$ is the position of an atom within the unit mesh. A diffraction function is now defined as F^2. It is this diffraction function which is the proportionality constant between the intensity and the interference function, i.e.

$$I(\mathbf{K}) = F^2(I.F.).$$

The importance of the diffraction function is that it alters the relative intensities of the scattered beams. Since F^2 is dependent on the atomic arrangement within the unit mesh, through the atomic positions $\mathbf{r}_n$, the comparison of intensities can be used to give structural information.

From the point of view of the Ewald diagram we may note that the interference function is related to the position of intersection of the reciprocal lattice rods with the Ewald sphere. The diffraction function is related to the thickness of the rods and thus to the area of intersection. The former factor defines the positions of the LEED spots, and the latter defines their intensities.

We now consider how the intensity of any particular spot may vary with electron energy. None of the above equations involve any explicit energy dependence, but in fact the atomic scattering factor f is weakly wavelength dependent. The result predicted is a slow uniform decline in intensity as energy increases. However, intensity profiles do not generally behave in this way. To account for the observations it is necessary to consider the contribution of lower layers—although still in the single-scattering approximation.

4. Kinematic theory including inner layers

As we have already seen, electrons can be expected to penetrate a few atom layers. It is reasonable to suppose, therefore, that the LEED pattern may also reflect these inner interactions. The argument proceeds by suggesting that the previously derived interference function, which includes the structural parameters **a** and **b** for the surface, be extended to include a contribution from the inner layers. Thus we can write

$$I.F = (I.F)_{surface} \times (I.F)_{inner\ layers}$$

and use the same function as before for the inner layers, i.e.

$$(I.F.)_{inner\ layer} = \frac{\sin^2(\frac{1}{2}M_3\mathbf{K} \cdot \mathbf{c})}{\sin^2(\frac{1}{2}\mathbf{K} \cdot \mathbf{c})}$$

where **c** is the unit mesh vector into the bulk.

This time, however, the number M_3 of scatterers is quite small, up to perhaps 5, since the electrons do not penetrate more than about five layers. The consequence of this low value of M is that $(I.F.)_{inner\ layer}$ does not significantly affect the positions of the spots. However, the inner layers do have a profound effect on the intensities of the spots. We can see, in principle, why this should be so by considering the two top layers of a crystal, as illustrated in Fig. 5.12. The angle θ_h at which a diffracted beam can be detected is determined, as before, by the conditions

$$\delta = a \sin \theta_h \quad \text{or} \quad \mathbf{K} \cdot \mathbf{a} = 2\pi h.$$

However, the wavelength λ' of the electron within the crystal differs from whether the electrons which penetrate the first layer and are scattered by the second layer emerge in phase or out of phase with the surface-scattered electrons. When they are in phase there is constructive interference and the intensity is a maximum; the opposite is true when they are out of phase.

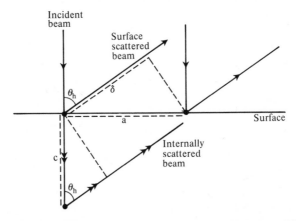

Fig. 5.12. Surface ($>$) and internally ($\gg$) scattered electron beams. The maximum reinforcement between the beams occurs when the path difference $c(1+\cos \theta_h)$ is an integral number of electron wavelengths.

In general terms Fig. 5.12 shows that the maximum intensity will be recorded when the path difference within the crystal Δ is an integral number of wavelengths, i.e.

$$\Delta = c(1 + \cos \theta_h) = \text{integer } (l) \times \text{wavelength of electron}$$

However, the wavelength λ' of the electron within the crystal differs from that in free space because the electron is now experiencing the inner potential V_0. The new wavelength is given by

$$\lambda' = \left(\frac{150}{V + V_0}\right)^{1/2}.$$

Now let us consider the effect on the $(0, 0)$ beam, for which $\theta_h = 0$, of the second-layer scattering. Constructive interference will occur when

$$2c = l \times \lambda'.$$

By substitution

$$2c = l\left(\frac{150}{V + V_0}\right)^{1/2}$$

In general, for planes of spacing d,

$$V = \frac{l^2 \times 150}{4d^2} - V_0.$$

A plot of the electron energies at which maxima in the scattered peaks are observed, known as the Bragg maxima, versus (integer)^2 is therefore a straight line of slope $(150/4d^2)$ and intercept $-V_0$. The experimental record of the intensity of a scattered beam as a function of the electron energy is known as an I–V plot or an intensity curve. As an example, the I–V plot for the $(0, 0)$ beam of cobalt (0001) is illustrated in Fig. 5.13 (the angle of incidence was slightly off normal, tipped by $7.7°$, to allow the reflected beam to be recorded). The Bragg peaks give the straight-line plot shown in Fig. 5.14, from which a value of the interlayer spacing in the surface region of 2.036 ± 0.006 Å was derived. This value is close to the bulk value of 2.046 Å. The inner potential had a typical value of 13.6 ± 1.7 eV.

Some observations on our discussion so far of the intensity profiles in LEED may be timely.

We note first that although the kinematic theory gave an adequate account of the positions of the intensity maxima in the I–V curve for cobalt (0001), no explanation was offered for the apparently haphazard variation of intensity from one peak to another. As it happens cobalt (0001) represents a rather favourable example for kinematic theory; in

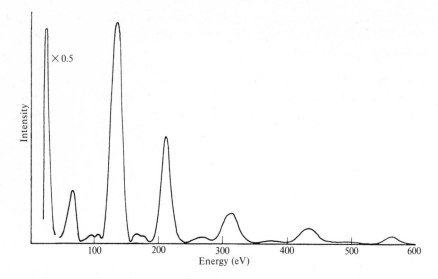

Fig. 5.13. *I–V* plot for the (00) beam of cobalt (0001) for the incident beam tipped 7.7° from the normal. The intensity is in arbitrary units. (From Benning, Allredge, and Viljoen 1981.)

other cases peaks additional to the Bragg peaks are recorded, for which the theory provides no explanation.

The fundamental reason for the complexity of *I–V* curves is that electrons interact strongly with matter. This is in contrast with X-rays whose interaction is much weaker, so that a kinematic, i.e. single scattering, approach is successful for X-ray diffraction. The consequence of

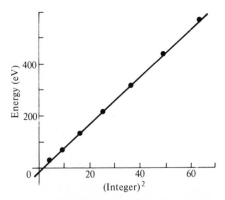

Fig. 5.14. Plot of the experimental values for the energy of the Bragg peaks versus (integer)^2. (From Benning *et al.* 1981.)

the strong electron/matter interaction is that multiple rather than single scattering is the characteristic behaviour. This multiple scattering both modulates the intensities of the Bragg peaks and introduces the possibility of 'non-Bragg-like' reflections. Although attempts to improve kinematic theory have been made they are inadequate and current practice is to use a multiple-scattering theory.

The second observation concerns the influence of an adsorbed layer on the intensity profile. As we have already seen, the spot positions are determined by the geometry of the surface layer, and indeed positions are independent of the scattering process. However, the interest in a surface layer is not merely for its geometry but also its atomic arrangement. That is to say, in the atomic positions of the adsorbed species on the surface. The contribution that intensity curves can make to the determination of atomic positions will be apparent if we refer again to Fig. 5.12. Suppose that the atom in the surface plane is an adsorbed species rather than the substrate. Then, in general, the new surface to first layer spacing, c', will be different from c. The electron path length within the solid is then different and the interference effect between surface-scattered and internally-scattered beams will be modified i.e. the intensity profile will be changed. Furthermore, if the adsorbed atom is located at some other atomic position altogether, say between the metal atoms rather than on top, a totally new intensity profile is to be expected. Correct analysis of this profile will give the atomic positions.

We now turn to the currently-favoured method of calculating I–V curves, which is known as a 'dynamical' method and involves the consideration of multiple scattering of the electron beam by the lattice.

5. Multiple Scattering or Dynamical Approach

This is essentially computational in character. However, a certain amount of insight into the physical principles behind multiple scattering can be obtained by considering a one-dimensional theory. This is largely artificial, in the sense that it will only apply accurately to either direct forward or direct back scattering and only for the low end of the energy spectrum. Within these limitations the one-dimensional approach does illustrate some general principles (Pendry 1974).

The model is of a regular array of scattering potentials, with a uniform background potential V_0 between each potential (Fig. 5.15). Each scatterer in one dimension is an analogue of a layer of ion cores in the real case. There is a step function at the surface to the vacuum level, which is taken as zero. In a multiple scattering model we can imagine waves travelling in each direction between the scattering centres as shown in Fig. 5.16. Every time a wave reaches a scatterer it is partly transmitted

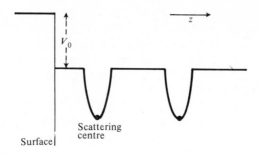

Fig. 5.15. One-dimensional potential for the dynamical theory of LEED.

and partly reflected. The contrast with kinematic theory arises from the fact that in this latter theory the wave is largely transmitted, reflection being weak. If now we consider the contributions to the forward scattered wave at a position X between two scatterers we can see that there are two contributions, as illustrated in Fig. 5.17.

1. There is the transmitted part of the positive wave moving between scatterers $n-1$ and n. This wave has an amplitude proportional to ta_{n-1}^+, where t is the probability of forward scattering, i.e. transmission.

2. There is the reflected part of the negative wave moving between scatterers $n+1$ and n. This wave has an amplitude proportional to ra_n^-, where r is the probability of backward scattering, i.e. reflection.

Thus in total:

$$a_n^+ \propto ta_{n-1}^+ + ra_n^-.$$

Similarly if we consider the reflected part of a_{n-1}^+ and the transmitted part of a_n^- we can deduce that the wave travelling in the reverse direction,

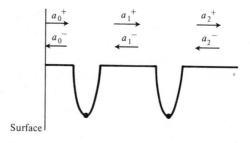

Fig. 5.16. Travelling waves in a multiple scattering one-dimensional model.

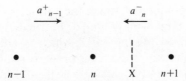

Fig. 5.17. Scattering of travelling waves at the nth centre. The amplitude at X in the forward direction is due to forward scattering a_{n-1}^+ plus back scattering a_n^-.

i.e. from scatterers n to $n-1$ has an amplitude a_{n-1}^- given by

$$a_{n-1}^- \propto r a_{n-1}^+ + t a_n^-.$$

Thus between each pair of scatterers there are two waves, one travelling in each direction, whose amplitudes are interrelated via t and r. The net result, two coupled waves, bears some resemblance to the motions encountered in classical physics for two pendulums which are coupled together, or to the use of normal modes in vibrational spectroscopy to analyse the motions of the atoms in a polyatomic molecule. Continuing this analogy leads to the result that the electron waves between scatterers have wavefunctions whose amplitudes vary by a constant factor each time the electron progresses from scatterer to scatterer, i.e. forward, $a_n^+ = \beta a_{n-1}^+$, and reverse $a_n^- = \beta a_{n-1}^-$. These are the normal modes. The equations apply for all values of n, except at the surface where the boundary condition that the wave inside the surface must match the incident wave is applied. The particular point of interest about the normal modes is that β is of the form $\exp(ikr)$. Thus, since the as have the periodicity of the lattice (by definition $\delta n = \pm 1$ is the move from one lattice point to its neighbour), the normal modes are mathematically equivalent to the Bloch waves mentioned in the discussion of the band model. Indeed the normal modes in LEED *are* the Bloch waves of a crystal.

Putting this conclusion another way, we note that in band theory the gaps occurred because of Bragg reflections within the metallic crystal, i.e. electrons were unable to propagate through the lattice. However, reflection of electrons, rather than propagation through the lattice, is precisely the condition for elastic reflection, i.e. the observation of LEED.

If we now turn to a three-dimensional treatment, the following elements of the theory can be identified.

1. Choice of the potential within the crystal with which the incident electrons interact. The muffin-tin potential described earlier is generally used.

2. Choice of the potential through the surface from the vacuum to the constant part of the muffin-tin potential. Fortunately, the scattering is not particularly sensitive to this potential which is not well known. As we have seen, the main effect of the electron on passing from the vacuum to the crystal is the change of wavelength to $\lambda' = \{150/(V + V_0)\}^{1/2}$.

3. The recognition that when interacting with the ion cores the electron is speeded up. As a result the wavelength is temporarily reduced and there is a shift in the phase of the scattered electron. The experimental consequence is a shift in the energy of the Bragg peaks. The value of the phase shift is an adjustable parameter which is available for fitting theory and experiment.

4. The requirement to take into account the fact that a substantial fraction of the electron scattering by the lattice is inelastic. This is a central feature of the electron–crystal interaction and is the fundamental reason why low energy electron diffraction is characteristic of just the top few atomic layers. In the Bloch-wave treatment described earlier (in the one-dimensional case) the inelastic scattering is taken into account by introducing an imaginary component into the inner potential experienced by the electron in the crystal. The potential is then written as

$$V_0 = V_{0_r} - V_{0_i}$$

where V_{0_r} is the real part of potential and V_{0_i} is the imaginary part.

Alternatively, and conveniently from a computation point of view, perturbation methods can be used to describe the scattering. The justification for this approach arises from the rapid decay in the electron beam as it penetrates the lattice. Rather few forward and back scattered intensities need to be considered to give rapidly converging intensity functions. Iterative procedures are then readily applied. The procedure is known as renormalized forward scattering. Also efficient in the use of computer time is the layer iteration method, which considers first intralayer scattering and then layer-by-layer scattering. (More details of the computational procedures, including programs, are to be found in Van Hove and Tong 1979).

As an example of the calculation of the intensity curve of a clean metal surface, let us return to the cobalt (0001) surface mentioned earlier. The computation was carried out using the renormalized forward scattering procedure. A range of top-layer spacings from 1.846 to 2.146 Å was investigated. The resulting calculated intensity curves, together with the experimentally observed curve, are shown in Fig. 5.18. As can be seen, there is good agreement at $d = 2.046$ Å, which is the bulk spacing.

Turning now to intensity profiles of the adsorbed layer, we take as an example CO adsorbed on nickel (001). The intensity profiles for three

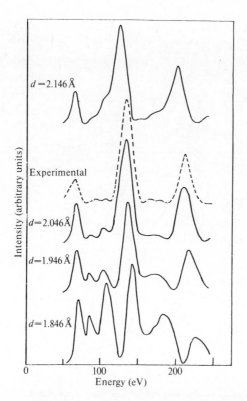

Fig. 5.18. Calculated *I–V* profiles for the (00) beams of cobalt (0001) at $\theta = 7.7°$.
- - -, Experimental curve. (From Benning *et al.* 1981.)

experiments are shown in Fig. 5.19, together with the theoretical profiles for the linear configuration Ni—C—O with the atomic spacings indicated. There is good agreement. However, it is worth emphasizing that LEED does not give atomic structures directly. On the contrary, it is necessary to assume a structure, calculate the associated LEED intensity profiles, and compare these with the experimental curves. Structure determination is therefore unlikely to become a matter of routine in the near future.

Finally on LEED we may note that theory has now advanced to the point where in some cases the reliability of results may be limited by the quality of the experimental data rather than the theory (Stevens and Russell 1981). For low index planes of metals that do not reconstruct, the reproducibility of intensity profiles between different instruments may be the limiting factor. At least part of the problem may arise from the need for the utmost care in degaussing the field-free region in which the

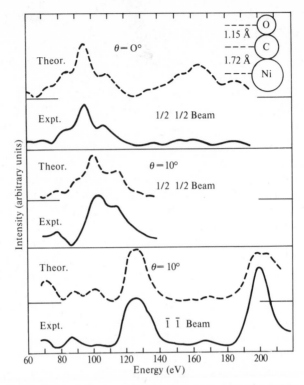

Fig. 5.19. Comparison of the experimental and theoretical LEED spectra from Ni{001}c(2×2)-CO. The structural model is shown schematically at the top. (From Passler, Ignatiev, Jona, Jepsen, and Marcus 1979).

electrons travel. Low energy electrons are readily deflected by magnetic fields so the relevant part of the chamber must be free from stray magnetic fields. A second point arises when weak beams, rather than the strong $(0, 0)$ beam we discussed above for cobalt (0001), are being recorded. A considerable accumulation time is then required during which even under ultrahigh vacuum conditions a significant amount of background gas (often CO) may adsorb on the crystal. Although regular flashing might seem to be the answer to the restoration of cleanliness, there may be problems in that the electrical perturbations caused by flashing can take a significant time to dissipate. The accumulation of data then becomes time consuming and places considerable demands on the stability of the LEED apparatus.

6. Electron emission

Introduction

The analysis of the energies and spatial distributions of electrons emitted from gases or solids has been an important source of information about the electronic properties of the emitting substance. Thus, for example, experimental verification of the band theory of metals (see Chapter 4), rests in part on the comparison of the measured intensity and energy distribution of electrons emitted under the stimulus of photon bombardment with the predictions of theoretical calculations of densities of states. In the study of adsorbed layers, electron emission has usually been brought about in one of two ways. The first is by bombardment with particles of energy appropriate to the electron it is desired to study. The second is by the application of a very high electric field (10^7–10^8 V cm^{-1}) which 'wrenches' electrons from the surface. Other possible methods of causing electron emission include heating, as in thermionic devices, and, occasionally, chemical reactions (exoemission). These latter methods are not of widespread application and will not be discussed further.

A variety of particles has been used as to initiate electron emission, the most commonly encountered being photons and electrons. Photon sources are usually in either the ultraviolet or the X-ray regions. A helium discharge lamp gives monochromatic ultraviolet (UV) radiation at 21.2 eV (He I) and 40.8 eV (He II), whilst Mg Kα or Al Kα are the preferred X-ray lines. Analysis of the ejected electrons gives an intensity which varies with electron energy, i.e. a spectrum. The technique is thus a branch of spectroscopy which is called photoelectron spectroscopy (PES). In the X-ray region the technique was originally called electron spectroscopy for chemical analysis (ESCA). However, when applied to the study of surfaces the self-consistent usage UPS to describe photoelectron spectroscopy at ultraviolet frequencies and XPS for the X-ray region has grown up. It may be mentioned that the energy gap between the UV and the X-ray regions has been bridged with the advent of synchrotron radiation. This is not as generally available as ultraviolet and X-ray radiation but is already providing valuable additional insights, especially into surface bonding. Electron beams are frequently used to observe the Auger processes mentioned in Chapter 4 which follow the ejection of a core electron. Electron energies of 1–5 kV are employed. Auger electron spectroscopy (AES) became widespread after it was appreciated that a LEED apparatus could readily be adapted for AES. It has now become standard practice to observe the Auger spectrum of metal surfaces being investigated by LEED to establish their initial freedom from impurity. A sensitivity to 1 per cent of a monolayer is claimed.

The techniques mentioned so far have been the source of a wealth of information about a wide variety of systems. One reason for this is that commercially available instruments often include facilities for LEED, AES, and either or both UPS and XPS. It is thus possible to investigate both structural and electronic properties of adsorbed layers in a single apparatus and under defined conditions. The results of some such investigations will be discussed later.

Other particles capable of causing emission include excited metastable neutral atoms, such as He(2^3S) (Penning ionization spectroscopy) and positive ions (ion neutralization spectroscopy). A summary of bombardment-induced electron emission processes is given in Table 6.1.

The alternative method of inducing emission of electrons is the application of an intense electric field ($\sim 10^7$ V cm^{-1}). The process is known as field emission. The required field is generated by shaping the specimen to a needle-sharp tip (radius <1000 Å) and subjecting it to a positive potential of several thousand volts. Under these conditions electrons escape from the surface by a quantum-mechanical tunnelling mechanism analogous to the proton tunnelling familiar in solution kinetics (Bell 1980). The obverse of field emission can also be observed by reversing

Table 6.1
Bombardment-induced electron emission processes

Name	Acronym	Short description
UV photoelectron spectroscopy	UPS	Source He I (21.2 eV) or He II (40.8 eV); valence electron energy analysed.
X-ray photoelectron spectroscopy	XPS (ESCA)	Source Mg Kα (1253.6 eV) or Al Kα (1486.6 eV); core electron energy analysed
Auger electron spectroscopy	AES	Secondary electron energy analysed following ionization of core by electrons or X-rays; monochromatic beam not required.
Ion neutralization spectroscopy[a]	INS	Auger electrons produced by positive ion bombardment
Penning ionization spectroscopy[a]	—	Metastable atoms cause electron emission

[a] Not discussed further.

the potential applied to the tip and admitting a low-molecular-weight gas (e.g. helium or hydrogen). The gas molecules thus come under the influence of the intense field when they approach the tip closely (i.e. within a few angströms). An electron from the gas may then tunnel to the metal and generate a positive ion. The process is called field ionization. Somewhat higher fields ($\sim 10^8$ V cm^{-1}) are needed than for field emission.

The experimental set-up is arranged so that the electrons emitted (in the first technique) or the ions produced (in the second technique) are accelerated by the electric field and collide with a fluorescent screen at a distance of a few centimetres from the metal tip. As a result, the particles produce a visible, magnified image of the metal tip. Because of this magnifying effect, the apparatus is a form of microscope and the techniques are called field emission microscopy (for electron emission) and field ion microscopy.

We shall now discuss the theory and some examples of the applications of various techniques mentioned above.

Ultraviolet photoelectron spectroscopy (UPS)

Let us start by considering the spectra of some simple molecules in the gas phase. We can then see how the spectra change on adsorption and how the changes give information about bonding at the surface. This is also the sequence which is followed in practice.

When a UV photon causes the ejection of an electron from a molecule, energy conservation requires that

$$E_{el} = h\nu - I - E_{vib} - E_{rot} \tag{6.1}$$

where E_{el} is the kinetic energy of the electron, $h\nu$ is the energy of the photon, and I is the ionization potential of the molecule. I is defined as the energy required to remove an electron from the molecule in its ground vibrational and rotational states and to produce a molecular ion also in these states; it is an 'adiabatic' ionization potential. E_{vib} and E_{rot} are the actual vibrational energies of the ion. Analysis of the energy of the electrons at fixed $h\nu$ (i.e. monochromatic radiation) thus provides values for the quantities on the right-hand side of eqn (6.1).

To date, the resolution of spectrometers has limited the observation of E_{rot} to the hydrogen molecule and this energy will not concern us further. However, the vibrational structure of simple molecules, that is to say the various values of E_{vib} with which the molecular ion may be produced, can be resolved.

To interpret the vibrational structure, potential energy diagrams for the molecule and the molecular ion are used. In general, we expect ionization of a diatomic molecule M to have one of three effects on the bonding. If a bonding electron is removed, M^+ is less strongly bonded than M. The potential energy well of M^+ is then shallower than that of M and its minimum is at a larger value of the internuclear separation. Removal of an antibonding electron would have precisely the opposite effect on these two properties. Ionization of a non-bonding electron would have little effect on the shape of the potential energy curve. These possibilities are illustrated in Fig. 6.1. As in optical spectroscopy, and in accordance with the Franck–Condon principle, the transitions are drawn vertically. The relative intensities of the emitted electrons are determined by overlap of the vibrational wavefunctions ψ_v, of M and M^+. Since $\psi_{v=0}$ has its maximum value at the centre whilst higher vibrational levels have maxima near the edges of the potential energy curve, the intensity patterns are as illustrated.

Now let us take the specific case of CO whose gas phase UPS spectrum is shown in Fig. 6.2. The relevant molecular orbitals in CO follow the sequence

$$(1\sigma)^2(2\sigma)^2(3\sigma)^2(4\sigma)^2(1\pi)^4(5\sigma)^2(2\pi)^0.$$

Of these orbitals, the 4σ is non-bonding and is localized on the oxygen atom whilst the 5σ is non-bonding and localized on the carbon atom. The $(4\sigma)^2$ and $(5\sigma)^2$ electrons are thus the lone pairs on oxygen and carbon respectively. The empty 2π orbital is antibonding. The bonding in CO is therefore associated with the $(3\sigma)^2$ and $(1\pi)^4$ electrons (the π orbital is

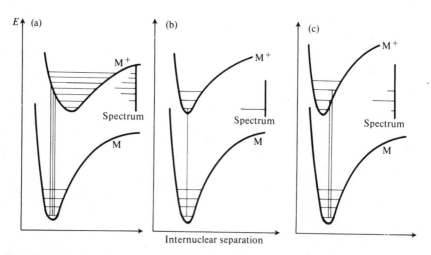

Fig. 6.1. Potential energy curves and related spectra for UPS. The electron removed in (a) is bonding; in (b) non-bonding; and in (c) antibonding.

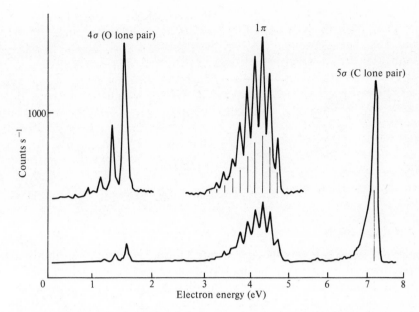

Fig. 6.2. UVPES of CO excited by HeI (21.2 eV) radiation. The upper curves are at higher recorder amplification. Calculated Franck–Condon factors are shown for the 1π band. (From Turner and May 1966.)

doubly degenerate). Approximate shapes of the molecular orbitals and the energy levels are shown in Figs. 6.3 and 6.4.

We can now interpret the CO spectrum shown in Fig. 6.2 in terms of the molecular orbitals. We note first that UV photons have sufficient energy to probe as far as the 4σ orbital so that the bonding 3σ orbital

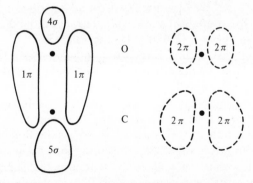

Fig. 6.3. Schematic representation of the molecular orbitals of CO. The 1π and 2π orbitals are doubly degenerate, with the additional orbitals lying above and below the plane of the paper. 4σ, 5σ, and 1π are occupied; 2π is vacant.

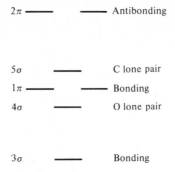

Fig. 6.4. Energy levels of CO (not to scale).

does not contribute to the spectrum. Turning next to the vibrational structure, we observe that the non-bonding 4σ and 5σ orbitals give peaks with little structure, whereas the bonding 1π orbital has considerable vibrational structure. All these observations are in accord with the principles illustrated in Fig. 6.1.

At this stage it is appropriate to consider a little more closely what property of the molecule it is that is measured in UPS. Part of the answer comes from a theorem due to Koopmans.

1. Koopmans's theorem

This theorem makes the simple statement regarding UPS that

$$\text{ionizational potential} = -\text{orbital energy}.$$

If this is correct, we have a direct experimental determination of the energy levels of the orbitals in a molecule. However, the theorem rests on a number of assumptions, in particular the following.

1. The electron distribution is the same in M^+ and in M, i.e. the other molecular orbitals are unperturbed by the loss of the electron. This is also known as the 'frozen orbital' approximation.

2. Electron correlation is the same in the ion and in the molecule.

3. Relativistic effects are the same in both M and M^+.

Although quantitative assessment of these assumptions is complex, it certainly seems intuitively likely that following ionization some relaxation of the remaining orbitals would occur to reduce the energy of M^+, i.e. assumption 1 is fairly obviously suspect. It may therefore be better to look on the theorem as a useful approximation to be tested by a comparison of theoretical and experimental orbital energies.

After this outline of gas phase UPS (for an excellent full account see Eland 1974), we turn to the more complex emission from metals.

2. UPS of metals

When a photon interacts with a metal, the limited number of well-defined energy levels characteristic of simple molecules is replaced by the band structure discussed in Chapter 4. A UV photon has sufficient energy to eject electrons from the bands due to valence electrons, although not from the atom cores. Since the energy spread within a band is about 10 eV, the ejected electrons are expected to have energies spread continuously over a similar range. From the point of view of adsorption and catalysis, it is the changes in emission brought about by the gas–metal interaction that are of interest. Such changes are normally recorded by subtracting the spectrum of the clean metal from that recorded following the formation of the adsorbed layer.

We can depict the process of photoemission from a metal on an energy level diagram, as shown in Fig. 6.5. One way of looking at photoemission is via the 'three-step model', in which it is broken down into the following steps.

1. Absorption of the photon at some arbitrary energy within the band by an electron of initial energy ε_i. This produces an excited electron of energy ε_f within the band. Energy conservation requires that $\varepsilon_f - \varepsilon_i = h\nu$.

2. Migration of this electron to the surface.

3. Escape of the electron.

The probability that excitation within the bond will occur in step 1 depends upon the wavefunctions associated with the initial and final states.

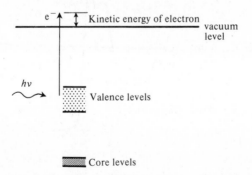

Fig. 6.5. Schematic representation of the ejection of an electron from a metal by a UV photon.

In calculations of probabilities, the Bloch functions mentioned in Chapter 4 are used. As far as the second step is concerned, we have already noticed in Chapter 5 that low energy electrons have a mean free path of about 10^{-9} m. Thus electrons ejected by UV photons will originate in the top few layers and UPS is expected to be a surface-sensitive technique. Although energy analysis of the ejected electron in the third step is straightforward experimentally, there may be some difficulty in deriving an accurate value for the energy of the level from which it has come. This difficulty is not obvious from the energy level diagram, which suggests that the kinetic energy is exactly determined by the work function of the metal and the particular energy level within the conduction band at which an electron originates. However, to be energy analysed the electrons must be collected; their recorded energies therefore depend on the 'effective' work function of the spectrometer. This quantity in turn depends on both the sample and the collector. Since the work function of the latter may be affected by extraneous factors, particularly adsorption, it is not well defined. Accordingly, it is usual to present photoelectron spectra of

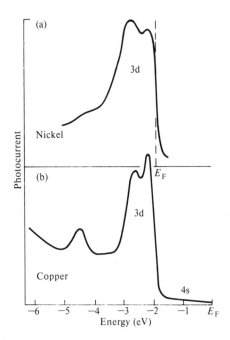

Fig. 6.6. Photoemission from (a) copper and (b) nickel (100) single-crystal planes. The energy scale for nickel has been shifted by 1.9 eV. (From Feuerbacher, Fitton, and Willis 1978.)

metals and adsorbed layers with energies measured relative to the Fermi level. This method of presentation is perfectly adequate for most purposes. When the work function is required, it can be obtained from measurements of the electron energy distribution.

As examples of photoelectron spectra of metals, Fig. 6.6 shows the results for copper and nickel. The contribution of 3d electrons at the Fermi level in nickel can be seen. In copper these electrons lie about 2 eV below the Fermi level, which is determined by the 4s band. These spectra reflect the band structure illustrated in Fig. 4.23.

Now that we have discussed the UV photoelectron spectra of gases and metals separately, it is time to consider what the effect of adsorption might be. Probably the most extensively studied adsorbed molecule is CO, whose gas phase spectrum was considered above. Although not all details of the spectra of CO adsorbed on metals are agreed, this molecule provides us with a good example of the way in which UPS has been used to deduce information about the changes in molecular energy levels, and thus of bonding, on adsorption.

3. UPS and adsorption: CO

The interpretation of spectral changes in the UPS spectrum of CO following adsorption relies considerably on evidence from the properties of the carbonyls of transition metals. In particular, the implications of synergic bonding for the relative energies of the molecular orbitals of adsorbed CO are central to the discussion. We start therefore with a brief account of this type of bonding (for more details see Cotton and Wilkinson 1980) in metal carbonyls.

CO bonds to metal atoms via the carbon lone pair $(5\sigma)^2$. A dative bond into an empty metal d orbital is thus formed. Simultaneously, occupied metal d orbitals back donate into the vacant antibonding (2π) orbital of CO. These processes are illustrated in Fig. 6.7. Since each electron shift

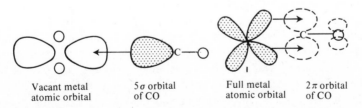

| Vacant metal atomic orbital | 5σ orbital of CO | Full metal atomic orbital | 2π orbital of CO |

Fig. 6.7. Schematic representation of the synergic bonding scheme in metal carbonyls.

tends to enhance the other, the overall effect is a cooperative strengthening of the bond, i.e. a synergic effect. In seeking experimental justification for the proposed bonding, it is natural to look first at the metal-to-carbon bond distances. It is to be presumed that the metal—carbon bond length is intermediate between that of a single and a double bond. However, data on metal—carbon bond lengths are sparse, so comparison of observed lengths with those predicted by consideration of other compounds is uncertain. Most experimental evidence in favour of the synergic bonding scheme has, therefore, come from consideration of changes in the C—O bond. Naturally, the length of this bond increases due to the occupancy of the antibonding (2π) orbital. Unfortunately, however, this length is not a particularly sensitive criterion of bonding, since the C—O distance does not vary greatly with bond order change between 3 and 2 (from (1.13 Å in CO(gas) to 1.15 Å in metal carbonyls).

The most weighty support for synergic bonding has come from infrared (IR) spectroscopy. The frequency of the C—O stretching mode depends upon the force constant of the C—O bond and thus on the shape of the potential energy curve for this bond at its minimum. The effect of back donation into the 2π antibonding orbital is to weaken the C—O bond and thus to decrease the IR stretching frequency (for more on IR spectroscopy see Chapter 7). In polynuclear metal carbonyls bonding to more than one metal atom may occur, since CO molecules occupy bridging positions between the metal atoms as well as the singly-bonded terminal positions. It is found that the more metal atoms to which the CO is bonded the greater is the reduction in the frequency. Experience has established a sequence of CO stretching frequencies in neutral carbonyls as follows: CO(gas), 2143 cm^{-1}; terminal CO, 2100–1850 cm^{-1}; double bridge-bonded CO, 1850–1750 cm^{-1}; triple bridge-bonded CO, 1730–1620 cm^{-1}. These characteristic frequencies play a major part in assigning vibrational spectra of adsorbed CO to particular types of surface bond, as discussed in Chapter 7. As far as UPS is concerned, the vibrational structure disappears on adsorption and the discussion centres on the shifts in the energies of the electrons. These shifts reflect the change in the orbital energies consequential on the formation of the surface bond.

Turning now to the UPS of adsorbed CO, we note first that this system has been and continues to be the subject of considerable activity, both theoretical and experimental. Some aspects of the bonding, e.g. whether the axis of the CO molecule is always perpendicular to the surface, have still to be resolved and detailed revision of bonding schemes is to be anticipated in the future. We shall follow the general line taken by Rhodin and Ertl (1979) and Brucker and Rhodin (1979). The first point to consider is the nature of the bond between the CO molecule and the

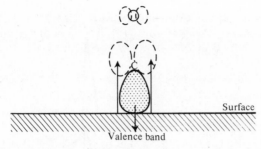

Fig. 6.8. Synergic bonding scheme applied to the adsorption of CO on a metal.

surface. A natural application of the synergic bonding discussed above leads to the scheme shown in Fig. 6.8.

First let us consider physical adsorption. This was defined in Chapter 1 as involving only van der Waals forces without any exchange of electrons, i.e. there is no synergic bonding. Experimentally, therefore, we should expect physisorbed CO to have a spectrum with peaks at energies little different from the gas phase. As an example, the UPS spectrum of about half a monolayer of CO physisorbed on copper at 20 K is shown in Fig. 6.9. As noted above, the Fermi level is taken as the reference point for the energy scale. For comparison, the UPS spectrum of gas phase CO is also illustrated. The predicted similarity between the peak energies of the adsorbed and gas phase spectra can be taken as corroborative evidence that there is only a weak interaction between gas and metal.

Turning now to chemisorption, we consider one possible interpretation of the effect of bonding on the energies of the CO valence orbitals. The most profound effect is to be anticipated for the 5σ orbital since this is converted from a non-bonding to a bonding orbital. Bonding has the effect of stabilizing an orbital, so we expect that the ionization potential of electrons from the new molecular orbital formed from the CO (5σ) plus a metal orbital will be significantly increased. The effect on the 1π and 4σ orbitals is more subtle. As we noted above, back donation into the antibonding 2π orbital stretches the C—O bond. This stretching affects the 1π and 4σ orbitals unequally; the π orbitals are more adversely affected, i.e. their energy is raised more, than the non-bonding 4σ orbital. Indeed, to a first approximation, this latter orbital may not be significantly affected. The stretching thus increases the energy of the 1π orbital both absolutely and relative to the 4σ orbital. An estimate of the differential energy shift between 4σ and 1π orbitals has put it at roughly 0.1 eV per 0.01 Å increase in bond length. Most of this effect is attributable to the 1π orbital energy change. Finally, if we consider the $1\pi - 5\sigma$

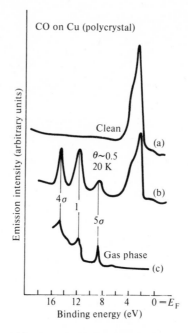

Fig. 6.9. UPS spectra of (a) clean polycrystalline copper, (b) copper covered with about half a monolayer of adsorbed CO, and (c) gas phase CO for comparison with (b). (From Brucker and Rhodin 1979.)

energy difference, this is decreased following adsorption since the 5σ orbital is stabilized and the 4π orbital is destabilized.

Some experiments which contributed to the foregoing picture concern the adsorption of CO on the iron (100) single-crystal plane. On the clean metal strong chemisorption is observed, whereas if the surface is pre-treated with sulphur, weak chemisorption occurs. The UPS results from CO adsorption on (a) clean iron (100) at 123 K and (b) sulphur-covered iron (100) at 98 K are shown in Fig. 6.10. As anticipated, weak

Fig. 6.10. (a) UPS results for strongly chemisorbed CO on iron (100): curve A, emission from clean iron (100); curve B, emission after CO chemisorption to an exposure of 3 L (1 L $= 1 \times 10^{-6}$ Torr s); curve C, difference spectrum B $-$ A; curve D, obtained after warming to 373 K showing the loss of emission from CO molecular orbitals and the appearance of the emission from the atomic orbitals O(1p) and C(2p). (b) UPS results for CO weakly chemisorbed on sulphur-covered iron (100): curve A, emission from sulphur-covered iron (100); curve B, emission from CO-covered S–Fe(100); curve C, difference spectrum B $-$ A. (From Brucker and Rhodin 1979.)

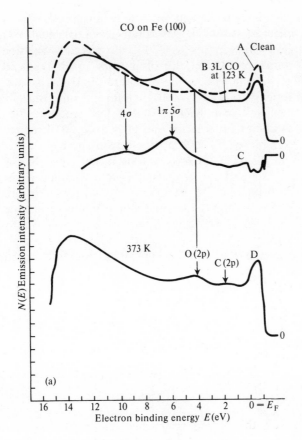

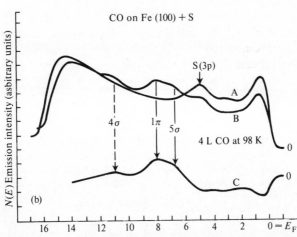

chemisorption causes the 5σ and 1π orbitals to approach in energy whilst the greater effect of strong chemisorption leads to actual overlap of the orbital energies. On warming to 373 K the electron emission from the molecular orbitals of CO-Fe(100) disappears and is replaced by peaks with energies characteristic of the atomic orbitals O(2p) and C(2p) (an inconspicuous peak). It is concluded, therefore, that on warming the bond stretching associated with back donation has proceeded to the point where the adsorbed CO molecule can be said to be dissociated.

Another example of the UV photoelectron spectrum of chemisorbed CO is shown in Fig. 6.11 for $\theta_{CO} = 0.6$ on nickel (100). Notice again the merging of the 5σ and 1π orbitals; this is a common observation for strongly chemisorbed CO. However, there is no reason in principle why the order of the 5σ and 1π should not actually reverse, i.e. for the electrons from the new molecular orbital derived from 5σ to appear at lower energy than the 1π. This has been suggested for the other low-index faces of nickel, the (100) and (111), although the assignment is far from straightforward.

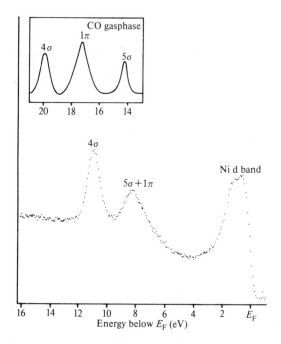

Fig. 6.11. UPS results for CO adsorbed on nickel (100) at $\theta \approx 0.6$. (From ref. 30 of Bradshaw 1979.)

X-ray photoelectron spectroscopy (XPS)

In general, the probability that a photon impinging on a sample causes an electron to be ejected from a particular energy level varies considerably with the photon energy. The form of the variation of this probability, as measured by the ionization cross-section, is illustrated in Fig. 6.12. Thus X-rays, whilst able to eject valence-level electrons, are more likely to cause ionization from atomic core levels.

The kinetic energy of an electron ejected from an atomic level by monochromatic radiation is well defined and is given by

$$E_{kin} = h\nu - \text{I.P.}$$

Since the energies of atomic core levels are usually peculiar to each atom, with comparatively few overlaps between different elements, XPS is in principle suitable for elemental analysis. Indeed that was an early important application and accounts for the original acronym ESCA (electron spectroscopy for chemical analysis). An indication of the electron energies to be expected in XPS can be gained if we note that for Mg Kα X-rays the photon energy is 1253.6 eV, whilst atomic core electrons have energies of some hundreds of electronvolts. Thus the energy scale for XPS is typically 100–1000 eV.

The energy range of XPS electrons is of some significance in determining which region of the sample is contributing to the observed signal. The escape depth of electrons from a metal (a difficult quantity to obtain reliably) varies with electron energy in the way shown in Fig. 6.13 (notice the log–log scale). Thus, as we have previously noted, electrons with energies in the range 10–200 eV, with which we are usually concerned in LEED and UPS, are at the minimum of the curve. The experimental observations in these techniques are thus sensitive to the interactions at or near the surface. On the other hand, electrons in the 1000 eV energy range can escape from numerous layers and they may therefore be more typical of the properties of the bulk.

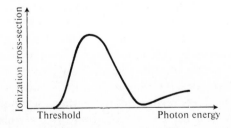

Fig. 6.12. Schematic representation of the dependence of the ionization cross-section on photon energy.

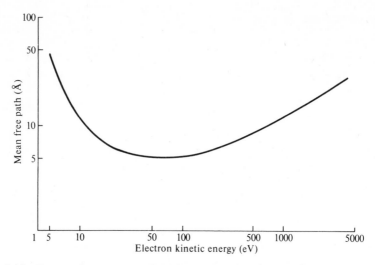

Fig. 6.13. General form of the energy dependence of the escape depth of excited electrons, showing the mean free path as a function of the kinetic energy.

There are two other factors which influence an X-ray photoelectron spectrum. The first is the observation that the ionization potential of an atomic core level depends to some extent on the chemical environment of the atom. The resulting shift in the energy of the ejected electron is known as the 'chemical shift'. This chemical shift, which is up to about 10 eV can give a valuable indication as to the type of bonding between the adsorbate and the surface. In particular, the position of the peak may be a diagnostic for deciding between molecular and dissociative adsorption.

As an example, let us continue the discussion of adsorbed CO on iron. In the XPS work the iron was polycrystalline and emission from the C(1s) orbital was monitored. Adsorption at 85 K gave a C(1s) peak at 285.5 eV, which is associated with the molecular adsorption. On warming to 350 K this peak disappeared and electrons with the lower energy of dissociated, or 'carbidic', carbon were recorded. Cooling the carbidized surface to 290 K in CO allowed further adsorption, now of molecular CO, to take place. The C(1s) spectra for this sequence are shown in Fig. 6.14, Two aspects of this work deserve comment. First, there is the satisfactory agreement with the UPS results quoted earlier. Secondly, we note that warming was required for CO dissociation. These considerations have been taken further in an interesting correlation made between XPS data (this time the position of the O(1s) XPS peak), the UPS peaks from CO

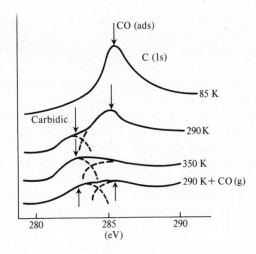

Fig. 6.14. XPS spectra of C(1s) from CO adsorbed on polycrystalline iron showing dissociation of CO(ad) on warming. (From Roberts 1977.)

molecular orbitals (($5\sigma + 1\pi$) and 4σ), and the heat of adsorption of CO. The result is shown in Fig. 6.15. As can be seen, a heat of adsorption greater than about 260 kJ mol^{-1} is expected to lead to dissociative rather than molecular adsorption. Iron, on which the heat of CO adsorption is about 180 kJ mol^{-1}, appears to be an example of a borderline case.

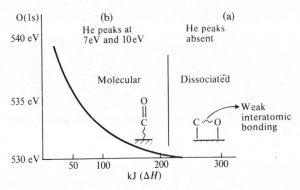

Fig. 6.15. Correlation between the spectroscopic and thermodynamic properties of adsorbed CO. (a) The position of the XPS peak from O(1s) is invariant, with an energy of 530 eV, for heats of adsorption greater than 260 kJ mol^{-1}. This value is typical of chemisorbed oxygen atoms on transition metals. (b) UPS peaks at about 7 eV and 10 eV below the Fermi level are characteristic of the ($5\sigma + 1\pi$) and 4σ orbitals respectively of chemisorbed CO in the molecular form. (From Roberts 1977.)

Although not dissociated during low temperature adsorption, warming CO-covered iron gives spectral features characteristic of the dissociated state.

The second factor to be considered is the occurrence of satellite peaks adjacent to the main peaks. The satellites are at lower electron kinetic energy, corresponding therefore to a higher ionization potential. These satellites are called 'shake-up' peaks. In using XPS for analysis, these peaks must be taken into account since they 'borrow' intensity from the parent peak. As well as the quantized transitions of the shake-up peaks, transitions to continuum states associated with the ion may also be possible. These are known as 'shake-off' peaks. Additionally, it should be noted that Auger electrons will often be recorded, as discussed below.

As with UPS, there may be some difficulty in determining accurately the absolute energies of the orbitals from which the electrons originate. Comparison with an agreed standard allows the results from different instruments to be compared.

Auger electron spectroscopy (AES)

When a core electron is ejected from an atom, the resulting ion is in a highly excited state. A number of possible processes may contribute to reducing the excitation. One of these is the emission of a secondary electron, the process being known as the Auger effect. Auger emission is caused when one electron drops from a higher level into the core vacancy, the energy liberated leading to the ejection of a second, or Auger, electron. An energy-level representation of this process, together with UPS and XPS for comparison, is given in Fig. 6.16. The energy of the secondary electron ejected in the Auger effect is analysed, as in XPS. As can be seen from Fig. 6.15 the energy of this electron is

$$E_{\text{Auger}} = E_z - E_x - E_y.$$

This equation shows that there is an important difference between AES and XPS in that the kinetic energy of the Auger electron is characteristic of the energy levels of the atom and independent of the energy of the exciting radiation. It is therefore not necessary to monochromatize the electron beam; this is an experimental convenience which makes electron-excited AES more popular (and cheaper) than X-ray excited AES. We have already noted that bombardment of a surface by an electron beam leads to the emission of secondary electrons. In AES, unlike LEED, these electrons cannot be rejected by suitable voltages applied to grids and are therefore detected. Since the Auger process is

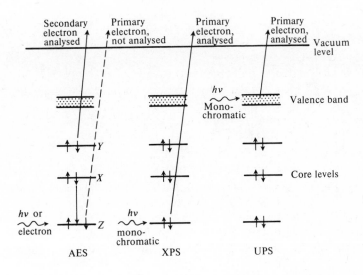

Fig. 6.16. Schematic summary of Auger, X-ray, and UV electron spectroscopies.

weak, it leads to an electron current which is minute compared with the current of secondary electrons. The quantitative detection of Auger electrons against this background emission, whose amplitude may vary considerably with energy, therefore poses problems. These are usually overcome by employing the electronic technique of phase sensitive detection (PSD). A consequence of using PSD is that the Auger peaks are recorded as the first derivative of the original curves and thus have the characteristic shape seen in Fig. 6.17.

A relaxation process which competes with Auger emission is X-ray fluorescence in which the energy liberated by the descent of the first electron into the core hole is liberated as a high energy photon. This process becomes more probable as the atomic weight increases, as shown in Fig. 6.18.

Although Fig. 6.16 shows the two electrons concerned in the Auger process as originating in two different atomic levels, this distribution is by no means essential. Other important possibilities are (i) that both electrons come from the same level and (ii) that one or both of them come from the valence levels. A system of nomenclature is clearly needed to identify the different possibilities. The convention has grown up that electrons originating in the (1s) shell are labelled K, the (2s) are L_1, the (2p) are L_2 and L_3, and so on. Valence shell electrons are called V. To

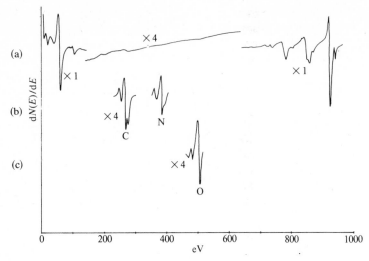

Fig. 6.17. Auger spectra of species adsorbed on a copper (111) single-crystal surface at 300 K: (a) clean copper (111); (b) after exposure to 9 L C_2N_2; (c) after exposure to 60 L O_2. (From Solymosi and Kiss 1981.)

identify a transition the following sequence of letters is used: 1, core hole; 2, hole generated by an electron dropping into the core hole; 3, hole generated by ejected electron. Thus in Fig. 6.16 the Auger electron would be described as (Z, X, Y).

1. AES and surface analysis

The energies of Auger electrons are clearly subject to the environmental factors which cause chemical shifts in XPS. Although of increasing impor-

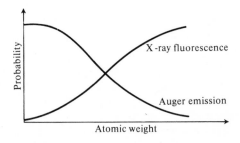

Fig. 6.18. Relative probabilities of X-ray fluorescence and Auger electron emission. The probabilities are equal at mass 34.

tance, comparatively little use has been made thus far of AES shifts, the overwhelming applications being to surface analysis. There are three main ways in which AES has been used.

The first is as an aid to assessing the cleanliness of metal surfaces prior to adsorption experiments. As previously noticed, this aspect has been most especially associated with LEED. A common impurity on metal surfaces is carbon, whose KLL spectrum is readily recorded. A sensitivity of about 1 per cent of a monolayer is claimed for AES, so that continuation of cleaning procedures until the carbon and other impurity peaks are absent ensures the initial cleanliness of the surface.

The second application of AES is to the measurement of the amount of gas adsorbed in a surface layer. For this purpose the peak-to-peak height of the AES derivative curve is a convenient measure of surface concentration. Calibration by some other technique, such as flash desorption, is required. As an example of this application, the Auger spectrum from cyanogen (C_2N_2) or oxygen adsorbed on copper (111) is shown in Fig. 6.17. The growth of the 384 eV signal of nitrogen and of the 271 eV signal of carbon (both due to KLL electrons) relative to the 920 eV signal of copper is shown in Figure 6.19. When the relative sensitivity of the instrument at 271 eV and 384 eV was taken into account, the carbon-to-nitrogen ratio was approximately 0.85, satisfactory close to the theoretical value of unity.

The third and most recent application of AES has been to the study of

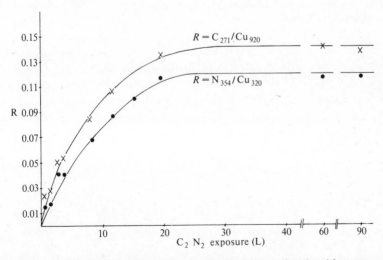

Fig. 6.19. Growth of relative carbon and nitrogen Auger signals with exposure to C_2N_2 at 300 K. (From Solymosi and Kiss 1981.)

the composition of the surfaces of alloys. As well as the intrinsic interest in this subject, there is the practical incentive of obtaining a better understanding of the surface properties of alloys since the introduction of supported bimetallic catalysts for re-forming reactions. The particular aspect of the properties of alloys with which we shall be concerned is the extent to which the composition of the surface layer deviates from the composition of the bulk. By way of example we shall discuss the Pt–Ir system.

The first point to be appreciated is that a thermodynamically ideal alloy is expected to have a surface composition different from the bulk, unless the heats of vaporization of the two metals happen to be equal. The basic theory of this surface enrichment of one component (for details see Williams and Nason 1974) considers the effects of mixing on the surface free energy. The bulk alloy is taken as an ideal solid solution. Then if the bond enthalpies are H_{AA}, H_{BB}, and H_{AB}, where A and B are the constituents of the alloy, we can write

$$H_{AB} = \tfrac{1}{2}(H_{AA} + H_{BB}).$$

The entropy of mixing associated with forming the surface layer is assumed to have the ideal value

$$S_{mix} = -k(X_A \ln X_A + X_B \ln X_B),$$

where X_A and X_B are the atom fractions. The calculation then proceeds to derive the composition of surface which minimizes the free energy. It is shown that the element with the lower heat of vapourization is enriched relative to the bulk. Calculation is required for each face individually since the extent of enrichment (or surface segregation) varies from face to face.

We can see qualitatively why the weaker lattice component tends to segregate at the surface by recourse to the 'broken-bond' model used in the calculations and illustrated in Fig. 6.20. As can be seen, formation of the surface generates 'broken bonds'. If now the 'guest' atom is of the less

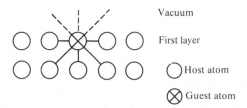

Fig. 6.20. Broken-bond model for alloy surfaces: ——, normal bonds; — — —, broken bonds.

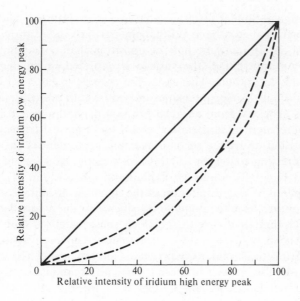

Fig. 6.21. Use of AES to measure the surface enrichment in a Pt–Ir alloy: — · —, experimental curve of the relative intensity $I_{Ir}/(I_{Ir}+I_{Pt})$ of the low energy peak due to iridium is plotted as a function of the relative intensity of the high energy iridium peak; — — —, predictions of the ideal solution model. (From Kuijers and Porec 1978.)

strongly bonded component, less energy is lost in forming the surface than if it is an atom of the more strongly bonded component. Hence the surface forms more readily, i.e. has a lower free energy, if it is concentrated in the weaker component.

The AES analysis of the Pt–Ir alloys makes ingenious use of the dependence of electron escape depth on energy shown in Fig. 6.21. By choosing to monitor two Auger transitions, one with energy near the minimum of the curve and the other with a high energy, signals reflecting the surface and bulk concentrations respectively are obtained. The energies of the transitions are 64 eV and 2044 eV for platinum and 54 eV and 1908 eV for iridium. After calibration, it can be shown that for iridium the fractional contribution I_1 of the first layer to the total signal I_∞ is

$$I_1/I_\infty = 0.44 \quad \text{at } 54 \text{ eV};$$

$$I_1/I_\infty = 0.08 \quad \text{at } 2044 \text{ eV}.$$

Thus nearly half (44%) of the electron current at 54 eV is due to the surface layer, whereas roughly 12 layers ($0.08 \times 12 \approx 1$) contribute to the

2044 eV signal. Similar results are obtained for platinum. The accuracy of AES is such that it is believed that the 2044 eV peak provides a better method of determining the bulk composition than conventional chemical analysis. As anticipated, the surface is depleted in the higher melting point metal, i.e. iridium. This result can be seen in Fig. 6.21 in which the surface and bulk Auger intensities due to iridium are compared with the intensities expected from an ideal solution of platinum and iridium. The general agreement is satisfactory, and if an empirical correction is made for a contribution by the second layer the agreement is excellent.

An interesting extension of this work using XPS has been made to supported bimetallic catalysts of platinum and iridium. These catalysts were prepared by coprecipitation of the metals from suitable salts onto an oxide support, usually alumina or silica, in finely divided form. An important point of difference between these catalysts and the samples used in AES is that the fraction of the surface covered by metal was less than 5 per cent. The intensity problems already associated with XPS are thus considerably greater in these cases, and accumulation times of many hours are needed. The plot in Fig. 6.22 for the XPS data shows, as expected, a deviation from linearity.

This concludes our discussion of bombardment-induced electron emission and we turn now to effects observed under the influence of large electric fields.

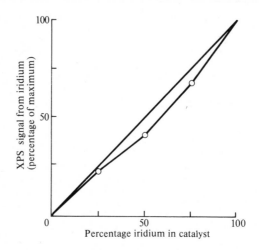

Fig. 6.22. XPS analysis of a Pt–Ir bimetallic catalyst. The iridium signal is expressed as a percentage: $\{Ir_{4f}/(Ir_{4f}+Pt_{4f})\}\times 100$. (G. L. Haller, unpublished work.)

Field emission and field ion microscopy

1. Field emission microscopy (FEM)

At room temperature and below a negligible number of electrons have sufficient energy to escape from the surface of a metal under normal circumstances. However, if a sufficiently large electric field is applied, electrons are able to leave the surface by quantum-mechanical tunnelling through the energy barrier represented by the work function. Let us now make an order of magnitude estimate of the field required for this process of field emission (for an authoritative account of field emission see Gomer 1961).

To a first approximation the application of an electric field to a metal gives rise to an energy diagram as shown in Fig. 6.23. For a current to flow, electrons must tunnel through a triangular barrier of height ϕ and width AB. In general, quantum-mechanical tunnelling becomes more probable as the barrier's height and width are reduced. However, the barrier height ϕ cannot be altered, so that to assist tunnelling the tunnelling distance AB must be reduced. A measurable current is expected when AB is less than about 15 Å. Since a typical value of ϕ would be 5 eV, the appropriate slope of BC, i.e. the electric field is about 5/15 V Å^{-1} or about 3×10^7 V cm^{-1}. A more detailed consideration shows that the barrier is rounded at the top, thus reducing the barrier height and increasing the current. A quantitative expression for the current i drawn from the specimen is given by the Fowler–Nordheim equation, which is of

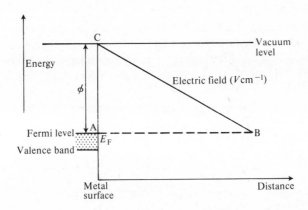

Fig. 6.23. Schematic energy diagram for the application of an electric field to a metal tip leading to field emission.

the form

$$\frac{i}{V^2} = A \exp\left(\frac{-b\phi^{3/2}}{V}\right),$$

where V is the voltage applied and A and b are essentially constant for a particular metal. A graph of $\ln(i/V^2)$ versus $1/V$ is known as a Fowler–Nordheim plot and its gradient is used to determine ϕ.

There are several interesting consequences for the study of surface layers of the Fowler–Nordheim equation. The first arises from the extreme sensitivity of the current to the work function. This latter quantity varies not only from metal to metal but also from one crystal plane to another on a particular metal. Thus, if emission from a variety of planes of a metal is studied under identical conditions, a range of electron currents is to be expected. Furthermore, if the adsorption of a gas causes a change in ϕ, as is commonly true, then the emitted currents will change considerably. These considerations illustrate the suitability of field emission as a technique for comparing the properties of adsorbed layers on different planes of a metal. The second point of interest is that the field-emitted current shows hardly any dependence on temperature—a few per cent only between 0 and 300 K. This makes field emission a convenient method of observing phenomena at low temperatures. Indeed, field emission experiments are frequently carried out at liquid nitrogen temperature (77 K) or below.

The usual method of observing field emission is by way of the field emission microscope. The principal features of this apparatus are shown in Fig. 6.24. The metal is etched to form a fine tip of radius <1000 Å and a positive potential of 1–5 kV is applied between the tip and the screen. Electrons then tunnel out of the tip and travel to the screen, a distance, typically, of 5 cm. This arrangement produces bright areas on the screen where the electrons hit it and gives a magnification of approximately $R/r \approx 5/10^{-5} = 5 \times 10^5$. It also generates the high field F required for field emission, since

$$F \approx \frac{V}{5r} = \frac{(1-5) \times 10^3}{5 \times 10^{-5}}$$

$$\approx (2-10) \times 10^7 \text{ V cm}^{-1}.$$

The resolution of the microscope is limited to about 20 Å by the lateral movement of the electrons.

By way of exemplifying the application of FEM we shall use some elegant studies of the adsorption and migration of oxygen and hydrogen on a tungsten tip. Because of the variation of the work function between

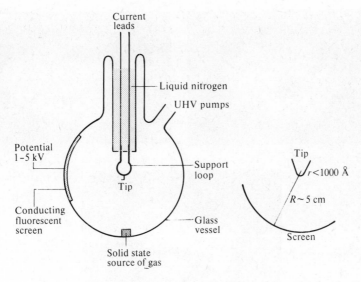

Fig. 6.24. Principal features of a field emission microscope. The screen is at a high (1–5 kV) positive potential. The solid state source of gas (ZrH_2 for hydrogen or CuO for oxygen) is heated electrically to release gas. When the microscope is immersed in liquid helium only the nearside of the tip becomes covered.

the planes exposed by the tip, a tungsten field emitter gives rise (ideally) to a four-leaf clover pattern, as illustrated in Fig. 6.25(a). On exposure to hydrogen or oxygen, adsorption caused the work function to increase and the emission to be reduced. Geometrically selective adsorption on the tip was achieved by locating the source of gas (ZrH_2 for hydrogen and CuO for oxygen) on one side of the globe containing the tip and immersing the whole microscope in liquid helium (see Fig. 6.24). Electrical heating of the solid state gas source generated the required gas, but only the side of the tip facing the source became covered; condensation of the gas on the cold walls protected the shadowed side from adsorption.

When the nearside of the tip was fully covered with chemisorbed oxygen, half of the pattern was blotted out, as illustrated in Fig. 6.25(b). In the first experiment the original exposure was more than sufficient to fill all the chemisorption sites, so a second layer of physisorbed gas was formed, i.e. $\theta > 1$. The situation was then as illustrated in Fig. 6.26. At liquid helium temperature (4.2 K) both the chemisorbed and the physisorbed layers were immobile. On warming (to about 27 K for oxygen or to <20 K for hydrogen) migration of the physisorbed, but not the chemisorbed, layer began. When the migrating physisorbed molecules

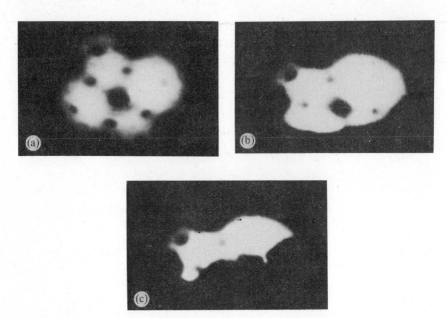

Fig. 6.25. The migration of oxygen on tungsten studied by field emission microscopy. (a) The FEM pattern from a clean tungsten tip; (b) the same tip, part of which is covered with adsorbed oxygen ($\theta_0 > 1$), at 4.2 K; (c) the same tip after 5.28 s at 27 K. (From Gomer and Hulm 1957.)

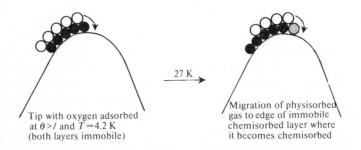

27 K

Tip with oxygen adsorbed
at $\theta > l$ and $T = 4.2$ K
(both layers immobile)

Migration of physisorbed
gas to edge of immobile
chemisorbed layer where
it becomes chemisorbed

Fig. 6.26. Schematic representation of the extension of a chemisorbed layer of oxygen on a tungsten tip at 27 K: ●, chemisorbed oxygen; ○, physisorbed oxygen; ◉, newly chemisorbed oxygen.

reached the edge of the chemisorbed layer they promptly adsorbed on the adjacent vacant sites. This process allowed further migration over the covered region and the extension of the chemisorbed layer and the dark area of the FEM pattern as illustrated in Fig. 6.25(c). rocess, which was likened to the unrolling of a carpet, continued a either all the physisorbed molecules had been used up, or the tip was fully covered with chemisorbed gas, depending on the initial coverage. The extension of the adsorbed layer is illustrated in Fig. 6.26.

Two deductions were made from these results: first, the chemisorption was non-activated since transfer from the physisorbed to the chemisorbed state was rapid at $\leqslant 27$ K. Secondly, from the rate of growth of the chemisorbed layer, an activation energy for surface diffusion of physisorbed oxygen molecules equal to about 4 kJ mol^{-1} was derived. If the tip was heated rapidly to 70 K, desorption from the physisorbed layer competed with migration and an activation energy for desorption of 12 kJ mol^{-1} was calculated. This energy can be equated with the heat of physisorption and compared with the latent heat of evaporation of oxygen which is 6.7 kJ mol^{-1}.

When the initial coverage of the exposed side of the tip was less than $\theta = 1$, i.e. no physisorbed gas was present, much higher temperatures were required to induce migration (180–240 K for hydrogen and 500–530 K for oxygen). The activation energy for this migration, E_{cm} (see Fig. 2.2) which was of chemisorbed atoms, was about 25 kJ mol^{-1} for hydrogen and 104 kJ mol^{-1} for oxygen. An additional point of interest is that these experiments show that, even at the top of the energy barrier to migration, the oxygen atom is still strongly bonded to the tungsten surface; since the strength of the W—O surface bond is about 630 kJ mol^{-1} at the top of the barrier the bond strength is $(630 - E_{cm})$, i.e. still about 526 kJ mol^{-1}.

2. Field ion microscopy (FIM)

The field ion microscope is essentially an FEM operated in reverse with a low-molecular-weight gas introduced into the microscope to generate the magnified image of the tip on the fluorescent screen. The ionization process occurs when the gas, usually helium or hydrogen at about 10^{-3} Torr, is within a few ångstroms of the surface. Following collision with the cold metal surface, the gas can be pictured as bouncing over it whilst becoming energy accommodated. During this time it is in a region of intense electric field and can be ionized by electron tunnelling into the metal. The resulting positive ion is accelerated to the fluorescent screen where its impact produces a dot of light. The magnification of the FIM is comparable with that of the FEM, but the resolution is somewhat better, being limited by the lateral motion of the imaging gas. This motion is less than that of electrons, and by operating at about 20 K a resolution of about 2.5 Å can be achieved. The FIM can thus resolve atomic dimensions, and field ion micrographs provide a striking visual record of atomic structure. An example is shown in Fig. 6.27.

One disadvantage of the FIM is that it exposes the tip to very large mechanical stresses ($\sim 10^{10}$–10^{11} dyn cm^{-2}). Only the strongest metals can

Fig. 6.27. FIM image of a tungsten tip.

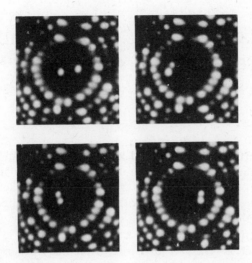

Figure 6.28. FIM images showing the migration of rhenium atoms over a tungsten
tip; at 375 K; diffusion interval about 3 s. (From Ehrlich 1980.)

survive and even for these mechanical fracture is a regular occurrence. However, under controlled conditions the field can be used to induce desorption of surface atoms, but stop short of rupture of the tip. This field desorption has been put to good use in two ways: first as a method of cleaning tips and secondly as a method of analysing surface species. For this latter use the adsorbed species, identified by a new spot on the micrograph, is field desorbed and analysed mass spectrometrically. This combination is known as the 'atom probe field ion microscope'.

Advances in the field of image intensification have allowed heavier inert gases than helium to be brought into use as the imaging gas. Their weaker light signals, compared with hydrogen or helium, are compensated for by image intensification. The advantage of heavier gases lies in the fact that lower electric fields are needed for field ionization. The range of metals which can be studied is thereby extended. Even with this advantage, only the most stable surface layers can be imaged. The most useful results from FIM are therefore probably those relating to adsorbed oxygen or to the migration of surface metal atoms, either over their own lattice or over a host. An example of this latter process is shown in Fig. 6.28.

7. Surface potentials and vibrational spectra

Introduction

The polar properties of an adsorbed layer provide a valuable source of information on the nature of bonding between adsorbent and adsorbate. The relevant properties are either (i) the permanent surface electric dipole moment associated with the formation of an adsorbed layer or (ii) the change in the dipole moment of a surface bond during vibration. The permanent electric dipole moment produces a change in the 'surface potential χ' which is reflected in the experimentally observed change in the work function ϕ of the surface. The importance of (ii), the change in surface dipole, lies in the general selection rule for a molecular vibration to be spectroscopically active in the infra-red (IR) region, i.e. that the dipole moment of the molecule must change during the vibration. This condition is usually expressed as:

$$\frac{d\mu}{dq} \neq 0$$

where μ is the dipole moment and q is the co-ordinate of the vibration. In the case of surface vibrations on metals this rule applies to motion perpendicular to the surface; modes parallel to the surface are inactive. Thus, when a surface vibration has $d\mu/dq \neq 0$ it is possible, in principle, to detect the surface species by IR spectroscopy. An alternative way of exciting the vibrational motion of adsorbed species is by electron impact using electrons with energies in the range 2–10 eV. Again, it is normally vibrations perpendicular to the surface that are excited, though it appears that for electrons this selection rule is not rigorous. Absorption of vibrational energy by the surface bond is recorded as a loss of energy by the electrons reflected from the surface.

Both for IR and electron energy loss spectra (EELS) the transitions are due to a change of one in the vibrational quantum number of the surface bond, i.e. $\Delta v = +1$. The rotational fine structure characteristic of gas phase IR is not seen for adsorbed molecules, since the interaction between surface and molecule blurs the rotational energy levels. Surface bands typically have a width $\geqslant 10$ cm^{-1} in the IR and $\geqslant 60$ cm^{-1} in EELS, this latter figure being the limit of instrumental resolution (a 'state-of-the-art' account of all methods available for studying surface vibrations is given in Bell and Hair (1980).

Surface potential measurements

It is helpful to an understanding of the electrical changes brought about by adsorption to consider first the distribution of the electrons at the surface of a clean metal. We begin by noting that one general condition of wave mechanics is that solutions of the Schrödinger equation, i.e. wavefunctions, shall be continuous. Thus, at the surface of a metal, although there is a sharp discontinuity in the atom cores, there is no such discontinuity in the electron distribution. Rather, the electron density dies away over a short distance into the vacuum. The result is a separation of charge, and therefore an electrical double layer, at the surface. This double layer can be modelled by a parallel plate condenser whose inner plate is at a potential Φ_{inner} and whose outer plate is at Φ_{outer}. Then, if an electron whose chemical potential in the bulk is μ_{el} is to escape into the vacuum, it must be provided with energy equal to the bulk potential plus enough energy to cross the surface double layer. By definition, this total energy is the work function ϕ, so we can write

$$\phi = -\frac{\mu_{el}}{e} + \Phi_{outer} - \Phi_{inner}.$$

The difference between Φ_{inner} and Φ_{outer} is the surface potential. Now μ_{el}, being a bulk property, is not affected by modifications to the surface layer, so changes in the work function due to adsorption are reflected only in changes in surface potential. Thus

$$-\phi = \frac{\mu_{el}}{e} + \chi$$

and

$$-\Delta\phi = \Delta\chi.$$

The significance of this relationship for adsorption comes from classical electrostatics, which relates $\Delta\chi$ to the dipole moment of the adsorbed layer by the equation

$$\Delta\chi = 4\pi\theta\mu n_{max} \tag{7.1}$$

where n_{max} is the number of molecules in a monolayer and μ is the dipole moment (assumed to be independent of θ) of each molecule in the layer. When the positive end of the surface dipole is away from the surface μ is positive and the work function is decreased by adsorption, and vice versa.

In practice, the assumption of a surface dipole moment which is independent of coverage is frequently unjustified and the linear relationship predicted by eqn (7.1) is not observed. Attention then focuses on two aspects of the measurements: (i) the values of $\Delta\chi$ and thus of μ as $\theta \to 0$, which gives information about the isolated surface—adsorbate bond and (ii) changes in the slope of the plot of $\Delta\chi$ as a function of θ. These latter changes may reflect the filling of new states on the surface.

1. Experimental methods

The change in surface potential resulting from adsorption, often more simply known as the 'surface potential', can be measured from changes in the work function. These changes are commonly obtained from the Fowler–Nordheim plots mentioned in the discussion of FEM in Chapter 6. Occasional use has also been made of changes in the photoelectric emission threshold following adsorption, using the energy equation for photoelectrons:

$$E = h(\nu - \nu_0)$$

where $h\nu$ is the energy of the photon and $h\nu_0$ is the threshold energy for photoemission, $e\phi$. A change in ϕ thus changes $h\nu_0$.

An alternative photoelectric method commonly employed when the apparatus includes UPS is to look at the shift in the low energy 'cut-off' in the spectrum. This effect is schematically illustrated in Fig. 7.1.

The change in the surface potential is frequently and conveniently measured by a condenser method. The metal of interest is made one plate of a parallel plate condenser, the other plate being of an inert metal such as gold. Because of the difference in work function between the two metals a potential difference, known as the contact potential, exists

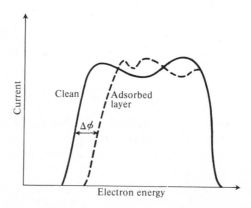

Fig. 7.1. Schematic representation of the change in the low energy cut-off of a UP spectrum due to a change $\Delta\phi$ in the work function following adsorption.

between them when they are connected through an external circuit. There is a consequential charge q on the condenser of capacity C given by

$$q = \text{contact potential}/C.$$

In the vibrating capacitor method, one plate vibrates at a suitable frequency, perhaps 500 Hz. Since C depends on the spacing between the plates the vibration generates a current in the external circuit connecting the plates ($i = \mathrm{d}q/\mathrm{d}t$). Application of a backing-off voltage opposed to the contact potential reduces the current to zero when the two voltages are equal. If adsorption causes a change in surface potential, a matching change in the backing-off voltage is required. Alternatively, in the static capacitor method, the tendency for current to flow in the circuit because of the change in contact potential when a gas is admitted is opposed by an automatically applied back-off voltage. In either case the zero-current balance point can be maintained throughout the experiment and a continuous record of variation of contact potential during adsorption is available.

Another convenient method for determining surface potential changes makes use of the characteristics of the thermionic diode. The flow of electrons from a heated cathode (such as a short length of tungsten wire) to an anode depends upon the applied potential in the way shown in Fig. 7.2. This characteristic curve shifts as shown when the surface potential of the anode is changed by adsorption of gas. The measurement is usually made when the anode is slightly positive and the diode is operating in the 'space-charge-limited' region. The technique is thus usually called the 'space-charge-limited diode'. The diode technique is often used in LEED apparatus, the electron gun being a convenient source of electrons. Although not strictly operating in the space-charge mode, curves similar

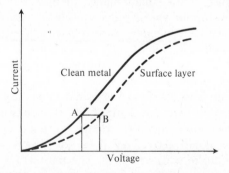

Fig. 7.2. General form of the characteristic curve of a thermionic diode. The voltage AB is the change in surface potential caused by adsorption.

to that in Fig. 7.2 are obtained. The method is straightforward and gives good results.

Finally, it should be noted that (a) these techniques are all based on the assumption that the work function of a reference device does not change during adsorption and (b) when the surface is not homogeneous the work function measured is an area average in all techniques, except the photoelectric threshold which measures the minimum ϕ.

2. Surface potentials and adsorption

Measurements of surface potential changes are currently seldom made as the sole technique for studying an adsorbed layer. Rather, they form one of a group of experiments which together allow a picture to be built up of the particular gas–solid interaction. In itself the sign of the change in the surface potential is a helpful criterion for discussing the bonding. Thus, for example, an increase in the surface potential (a decrease in the work function) implies that the electron shift is from the adsorbed species to the metal. Such a shift is to be anticipated for easily ionized metals adsorbed on metals of high work function. Determination of the surface dipole moment, which requires a knowledge of the absolute coverage, may quantify the extent of the electron transfer and thus indicate whether bonding is essentially ionic or covalent.

An example where ionic bonding is indicated is, unsurprisingly, in the adsorption of alkali metals on tungsten. Thus, for caesium the surface dipole, at zero coverage, is about 8 debye, a high value which agrees with that calculated for full electron transfer and is comparable with the gas phase dipole moment of the CsF molecule. It is interesting to note in passing that the formation of ions by alkali metals adsorbed on tungsten is made use of in molecular beam studies. The flux of alkali metal atoms in a beam is measured by allowing the beam to impinge on a tungsten wire hot enough to desorb the alkali metals (about 2000 K). A significant fraction of the desorbed species are the alkali metal ions, which are collected by a cylinder concentric with the wire. The measured current carried by the positive ions is directly proportional to the flux.

Adsorption of the reactive gases nitrogen, oxygen, or CO leads to smaller changes in surface potential which are of opposite sign from those of the alkali metals. This result is interpreted as due to the formation of a negatively charged layer of covalently bonded chemisorbed species. The variation of surface potential with coverage is often of some complexity—that is to say, as noted above, $\Delta\chi$ does not always vary linearly with coverage in accordance with eqn (7.1). The reasons for the non-linearity are varied and each example needs individual consideration.

The case of hydrogen adsorption on palladium (100) provides a good

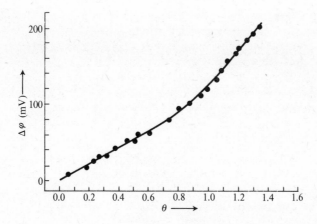

Fig. 7.3. Variation of the work function change with coverage for hydrogen on palladium (100). (From Behm, Christmann, and Ertl 1980.)

example of the application of a group of techniques, including work function measurements, to an adsorption system. The work function change was measured by the vibrating-electrode method and was found to be directly proportional to the coverage of hydrogen up to $\theta = 0.9$. Above this coverage the rate of increase of ϕ with θ increased. The plot of $\Delta\phi$ versus θ is shown in Fig. 7.3. The reason for the change of slope at

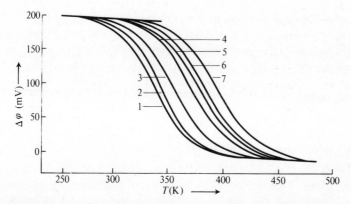

Fig. 7.4. Adsorption isobars for hydrogen on palladium (100) measured by work function changes at various pressures (in Torr): curve 1, 3×10^{-9}; curve 2, 5.3×10^{-9}; curve 3, 1.8×10^{-8}; curve 4, 8.5×10^{-8}; curve 5, 1.4×10^{-7}; curve 6, 2.5×10^{-7}; curve 7, 5×10^{-7}. (From Behm *et al.* 1980.)

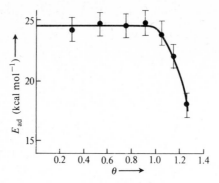

Fig. 7.5. Isosteric heat of adsorption of hydrogen on palladium (100) as a function of coverage. (From Behm *et al.* 1980.)

$\theta = 0.9$ was the formation of a new state on the surface, a conclusion confirmed by the appearance of an extra feature in the thermal desorption spectrum at $\theta > 0.9$ (The value of $\theta > 1$ seen on Fig. 7.3 implies that there is more than one surface hydrogen atom per surface palladium atom; the maximum uptake was $\theta = 1.35$, corresponding to $n_H = 1.8 \times 10^{15}$ cm^{-2}). Additional information about the Pd—H bonding was obtained by using the work function change to monitor the temperature dependence of the surface coverage, at equilibrium hydrogen pressures between 3×10^{-9} and 5×10^{-7} Torr. The adsorption isobars are shown in Fig. 7.4. Calculation of the isosteric heat of adsorption as described in Chapter 1, allowed ΔH_{ad} to be calculated as a function of θ. The result is shown in Fig. 7.5. The abrupt decline in ΔH_{ad} above $\theta = 0.9$ correlates well with the two other changes at this coverage just discussed. Other measurements included the sticking probability curve which had the typical precursor shape and the LEED patterns which showed a c(2×2) structure at $\theta = \frac{1}{2}$.

Infra-red spectroscopy of adsorbed layers

There are several ways in which IR spectroscopy has been used to study adsorbed layers. These include the familiar transmission method used for gases and solutions, the application of Fourier transform techniques, IR ellipsometry and reflection–absorption IR. The most commonly used technique is transmission IR and we shall discuss this, together with an example of its application, in more detail than the other techniques

1. Transmission IR and its application to CO on rhodium

In this experiment the sample is in the form of small metal crystallites deposited on a finely divided inert oxide support. Careful choice of conditions is required since metals are strong absorbers of IR radiation and the support particles tend to scatter the radiation unless their diameter is small compared with the wavelength. Scattering is also promoted by loose packing of the particles. Spectra are optimized, therefore, by using a metal coverage of perhaps 2–5 per cent on support particles of diameter ~200 Å, and compressing the particles in a die to form a thin disc. This disc is mounted in a vacuum, or sometimes an ultra-high vacuum, cell with windows transparent to IR radiation. The experimental gas is then admitted to a chosen pressure. This cell is mounted in one beam of an ordinary double-beam spectrometer. The spectrum of the support and metal after careful cleaning is recorded and subtracted from subsequent spectra. It should be noted that the support places limitations on the spectral regions which may be studied. In the first place it may have strong absorption regions which effectively 'black-out' a portion of the spectrum; secondly transmission at frequencies less than about $800 \, \text{cm}^{-1}$ is usually inadequate.

Although a wide range of adsorbed gases has been studied, a favourite molecule is CO which has a strong IR absorption band in a convenient region of the spectrum. To illustrate the use of transmission IR spectroscopy we shall consider the adsorption of CO on rhodium supported on high area alumina. A minor difference in this experiment from that described above is that instead of making a mechanically self-supporting disc of the sample, the CaF_2 window of the IR cell acts as the support. The experimental cell, which could be baked and pumped to a minimum pressure of about 1×10^{-8} Torr, is illustrated in Fig. 7.6.

The specimen was prepared by stirring a high area alumina powder into a solution of $RhCl_3$ in an acetone–water mixture. The resulting slurry was sprayed onto one CaF_2 window demounted from the cell and maintained at 80 °C. Rapid evaporation of the solvent left an adherent coating on the window. After remounting the window, the $RhCl_3$ was reduced to metallic rhodium by heating sequentially *in vacuo*, in hydrogen (three times) and again *in vacuo*. Spectra were then recorded at 295 K for CO pressures up to about 50 Torr. In addition, the adsorption isotherm up to $p \approx 1$ Torr was measured to allow the spectral intensities, obtained by integration of the IR peaks, to be expressed in terms of CO uptake by the metal. A linear relationship between peak intensity and adsorbed gas was found. Slow equilibration prohibited isotherm measurements for $p > 1$ Torr. The maximum uptake at 50 Torr of CO corresponded to a ratio of adsorbed CO molecules to surface rhodium atoms of $N_{CO}/N_{Rh} = 0.92$.

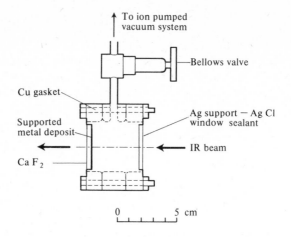

Fig. 7.6. Side view of a cell for transmission IR of CO on Rh/Al₂O₃. (From Yates, Duncan, Worley, and Vaughan 1979.)

The most important results were the coverage-dependent spectra shown in Fig. 7.7. As already noted in Chapter 7 when interpreting UPS spectra, IR absorption at 1600–2100 cm^{-1} is associated with the C—O stretching mode. Accordingly, all the peaks in Fig. 7.7 are assigned to this mode. The results can be summarized as follows.

1. There is a broad band (band 1) which is centred at about 1866 cm^{-1} at low coverage and shifts up in frequency to 1870 cm^{-1} at maximum uptake.

(2) There is a weak band (band 2) which shifts up in frequency from about 2050 cm^{-1} to 2070 cm^{-1} as the coverage increases.

3. The most pronounced feature is an intense doublet (band 3) with peaks at 2101 cm^{-1} and 2031 cm^{-1} which do not shift with coverage. The relative intensity of the components of the doublet is independent of coverage.

In discussing these results the first point to be noted is the upward shift in frequency with coverage of two bands. Such a shift is frequently, though not universally, observed for CO adsorbed on transition metals. One way of explaining this shift is by application of the synergic bonding model illustrated in Fig. 6.8. As the coverage increases, the competition between the CO molecules for the metal d electrons may reduce the extent of the back donation into the 2π (i.e. antibonding) molecular orbital of individual CO molecules. The bond-weakening effect of the

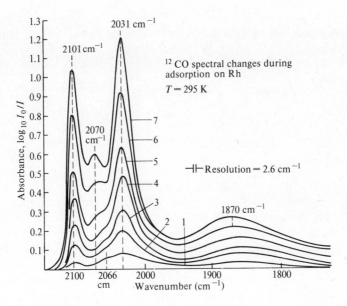

Fig. 7.7. IR spectra for ^{12}CO adsorbed on rhodium for increasing CO coverage ($T = 295$ K): curve 1, $p_{CO} = 2.9 \times 10^{-3}$ Torr; curve 2, $p_{CO} = 4.3 \times 10^{-3}$ Torr; curve 3, $p_{CO} = 5.0 \times 10^{-3}$ Torr; curve 4, $p_{CO} = 8.3 \times 10^{-3}$ Torr; curve 5, $p_{CO} = 0.76$ Torr; curve 6, $p_{CO} = 9.4$ Torr; curve 7, $p_{CO} \approx 50$ Torr. (From Yates *et al.* 1979.)

synergic effect on the C—O bond thus diminishes and the vibration frequency increases. Isotopic studies using ^{13}CO showed that this 'chemical' effect was not the whole story, because there were different shifts of ^{12}CO and ^{13}CO when mixtures were adsorbed on supported platinum. It is now thought that three 'physical' mechanisms may be operative between adsorbed CO molecules. These are a dipole–dipole coupling between adjacent vibrating molecules in which the oscillation of one surface dipole affects the potential energy and thus the vibration frequency of its neighbour. This effect is much greater for like molecules whose frequencies match than for unlike molecules, even isotopic ones such as ^{12}CO and ^{13}CO. The second suggestion is that the coupling between the surface dipoles may include a contribution from a mechanism involving vibrational interaction via the conduction electrons of the metal. The third effect is a direct intermolecular repulsion between the surface dipoles. The quantitative assessment of the relative importance of these effects is not settled and continues to engage interest (for a full account of vibrational shifts of CO on copper see Chapter 4 of Bell and Hair (1980)). As far as the interpretation of the results for CO on rhodium is concerned, the most significant aspect of theories of IR shifts is that they are expected to apply more particularly to polyatomic clusters of atoms than to individual isolated metal atoms.

The assignment of the IR bonds has been attributed to the following surface species:

Bond number: 1   2   3

The justifications for these assignments can be given as follows. First, the coverage-dependent shifts for bands 1 and 2 indicate that the metal particles contain the group of atoms required for such shifts to occur. Secondly, the coverage-independent frequencies of the doublet is strongly indicative that the rhodium atoms associated with structure 3 are in fact *isolated* rhodium atoms on the Al_2O_3 support'. The assignments are further supported by comparison with rhodium carbonyls. Thus, the halogen bridge-bonded compounds ($X \equiv Cl$ or Br) give rise to doublet

bands with frequencies 2093 ± 2 cm^{-1} and 2042 cm^{-1}, which are closely comparable with the values of 2101 cm^{-1} and 2031 cm^{-1} assigned to structure 3. The modes for the doublet structure are the coupled symmetric and antisymmetric C—O stretching vibrations of both adsorbed CO molecules. Furthermore, structures 1 and 2 are in accord with the discussion in Chapter 6, the larger shift to lower frequencies being associated with the bridge-bonded structure 1.

Other deductions, made from a more detailed analysis of the results, were that the fraction of the surface sites present in the form of the individual atoms giving rise to structure 3 was in the range 30–60 per cent. It was also concluded that the bonding angle in structure 3 is near 90°.

2. Other IR techniques

The application of Fourier transform (FT) techniques to spectroscopy has produced a great advance in the study of systems giving weak signals. Nowhere is this more noticeable than in modern nuclear magnetic resonance

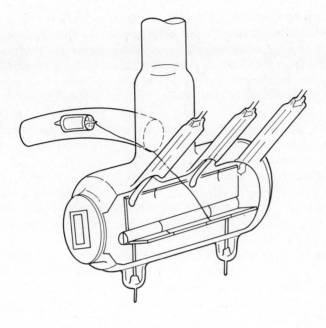

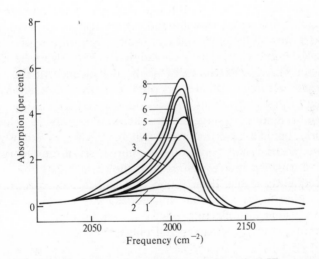

Fig. 7.8. (a) Apparatus for measuring reflectance–adsorption IR spectra. The copper film was deposited onto the hinged glass support plates (shown in the open position) which were then closed to an accurately parallel position for absorption measurements. (b) Spectra of CO on the copper mirrors at room temperature at the following pressures (in $N m^{-2}$): curve 1, 0.12; curve 2, 0.40; curve 3, 1.3; curve 4, 5.3; curve 5, 13; curve 6, 67; curve 7, 210; curve 8, 900. (From Pritchard and Sims 1970.)

spectroscopy where it is now routinely possible to observe ^{13}C spectra using the natural abundance (1 per cent) of [13]C. In the IR region the advantage of an FT spectrometer over a grating instrument, such as that used for the work on CO on rhodium described above, is about sixteen fold in the CO stretching region. However, this advantage is obtained at the expense of considerably greater experimental sophistication. The FT instrument comes into its own when the time of observation is of the essence. Thus, if many samples have to be recorded or if the spectra are liable to change over the period of observation, the more rapid response of FT techniques becomes especially valuable.

In IR ellipsometry, a beam of IR radiation is reflected specularly (i.e. $\theta_{reflected} = \theta_{incident}$) from a metal surface. The optical characteristics of the surface, that is to say its reflection coefficients for polarized radiation, change following adsorption of gas. The spectrum from the adsorbed gas is analogous to a normal IR spectrum. However, the significant advantage of this technique over normal IR is that the signal observed depends only on the optical properties of the surface. It can therefore be used at a high pressure of ambient gas and thus under conditions similar to those obtaining in practical catalytic processes; normal IR is not suitable for these conditions.

Finally, we note an important advance in the application of IR spectroscopy to adsorbed layers through the reflectance–adsorption method. This was developed following the theoretical prediction that a layer adsorbed on a bulk metal surface, rather than on a supported metal as previously studied, could be observed by IR. The key to the success of the experiment lay in the choice of angle of incidence of the radiation. A sharp peak in the absorbance was predicted at an angle 88° to the surface normal. At this angle about 30 reflections were predicted to be enough to produce a readily observed absorbance. The way in which multiple reflections on an evaporated metal film were achieved is illustrated in Fig. 7.8. The model system chosen for the initial test of the theory was CO on copper, for which the results are also shown in Fig. 7.8.

Subsequently it has been found that satisfactory spectra can be obtained with a single reflection, which does away with the necessity for the accurately aligned reflecting plates shown in Fig. 7.8. It also makes possible the study of single-crystal planes by IR.

Electron energy loss spectroscopy (EELS)

Excitation of the vibrational motion of surface species by electrons corresponds to a loss of energy up to about 400 meV (1 meV = 8 cm^{-1}). To observe this energy loss, which is a form of inelastic scattering, a well-monochromatized beam of electrons impinges on a metal single-crystal plane and the specularly reflected beam is detected. Energy analysis of the electrons in the reflected beam provides the energy-loss spectrum. In a typical experiment (for details see Chapter 4 of Bell and Hair (1980)) the initial beam current is 0.1–1 nA of electrons with energy in the range 2–10 eV. The beam impinges at an angle of 70° to the surface normal. The specular elastically scattered beam is of intensity 10^5–10^6 electrons s^{-1}, whilst the inelastic peaks (i.e. the spectrum) have intensities of 10–1000 electrons s^{-1}.

The experimental conditions place stringent requirements on the design of the apparatus. Electron beams of such low energy are readily deflected by magnetic fields, so excellent magnetic shielding is essential. The sparse electron fluxes in the spectral peaks require counting techniques to be used for detection, whilst the requirements of surface cleanliness necessitate ultra-high vacuum conditions for the whole spectrometer. A schematic representation of a high resolution EEL spectrometer is shown in Fig. 7.9.

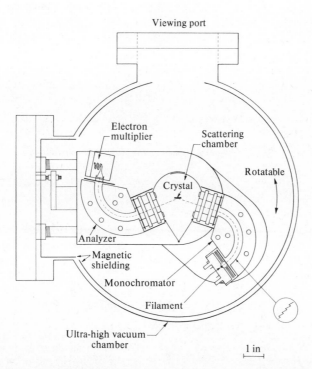

Fig. 7.9. Schematic representation of a high resolution electron energy loss spectrometer. (From Dubois and Somorjai 1980.)

Once again adsorbed CO has played a major role in the development of this new surface science technique. For an example of the use of EELS we shall continue with our discussion of the adsorption of CO on rhodium, this time the single-crystal rhodium (111) face. The EEL spectrum was recorded both as a function of exposure to CO and under equilibrium conditions. The pressure of CO was in the range 5×10^{-8}–1×10^{-5} Torr. The spectra are shown in Figs. 7.10 and 7.11.

The salient features of Fig. 7.10 are as follows:

1. the growth and shift to higher frequency of the principal peak, initially at 1990 cm^{-1};

2. the growth and shift to lower frequency of the peak first observed at 480 cm^{-1};

3. the appearance of a shoulder on the main peak at an exposure between 0.4 and 1.0 L.

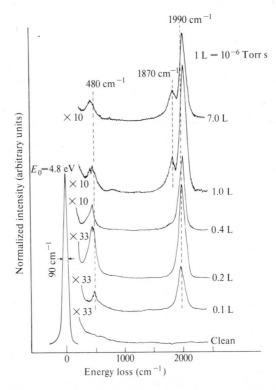

Fig. 7.10. Vibrational spectra of CO chemisorbed on an initially clean rhodium (111) single-crystal surface at 300 K as a function of gas exposure. Note the shift in both the 480 and 1990 cm^{-1} losses with increasing surface coverage. (From Dubois and Somorjai 1980.)

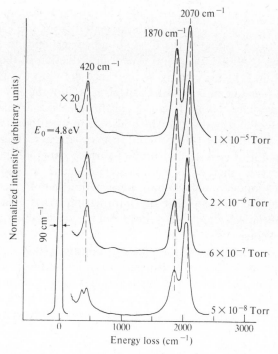

Fig. 7.11. Vibrational spectra of CO chemisorbed on rhodium (111) at 300 K as a function of background gas pressure. The loss above 2000 cm^{-1} reaches a limiting value of 2070 cm^{-1}, while the peak at 1870 cm^{-1} increases in intensity at a constant frequency. (From Dubois and Somorjai 1980.)

Turning now to Fig. 7.11 we note the following:

1. the highest frequency peak continues to shift to higher frequency, reaching a limiting value of 2060–2070 cm^{-1};

2. the low frequency peak shifts further to 420 cm^{-1};

3. the shoulder at 1870 cm^{-1}, whilst growing substantially, does not shift.

The following interpretation of these results was given.

1. the highest frequency peak 1, shifting from 1990 to 2070 cm^{-1}, is the C—O stretch of linear bonded CO;

2. the lowest frequency peak 2, which shifts from 480 to 420 cm^{-1}, is the Rh—C stretch;

3. the shoulder 3, which grows into a substantial peak at the highest CO pressure, is due to the C—O stretch of bridge-bond CO.

The CO is thought to be bonded perpendicular to the surface since no bending modes were recorded. The upward shift in the frequency of the linear-bonded CO is accompanied by a downward shift in the Rh—C frequency. This result is entirely as anticipated from the earlier discussion, since a strengthening of the C—O bond corresponds to a weakening of the bonding to the surface, and thus to a reduction in the strength of the Rh—C bond. The only point of difference between the IR data from the supported metal and the EEL spectra of the single crystal is the absence of the slight (about $15\,cm^{-1}$) shift in the bridge-bonded peak at $1870\,cm^{-1}$. Overall, the results, which also fit in with thermal desorption peak coverage shifts, must be counted in excellent agreement.

From the point of view of the study of surface structure, breakdown of the selection rule that only modes perpendicular to surface can be observed is advantageous since more information thereby becomes available. An interesting example of the information to be gained from the breakdown of the selection rule comes from the study of hydrogen on the tungsten (100) face.

At saturation, the hydrogen uptake on tungsten (100) is 2×10^{15} atom cm^{-2}, corresponding to two hydrogen atoms per surface tungsten atom. The arrangement of hydrogen atoms is thought to be the bridge structure illustrated in Fig. 7.12. The vibration perpendicular to the surface gives a strong allowed absorption feature at 130 meV. However, by making

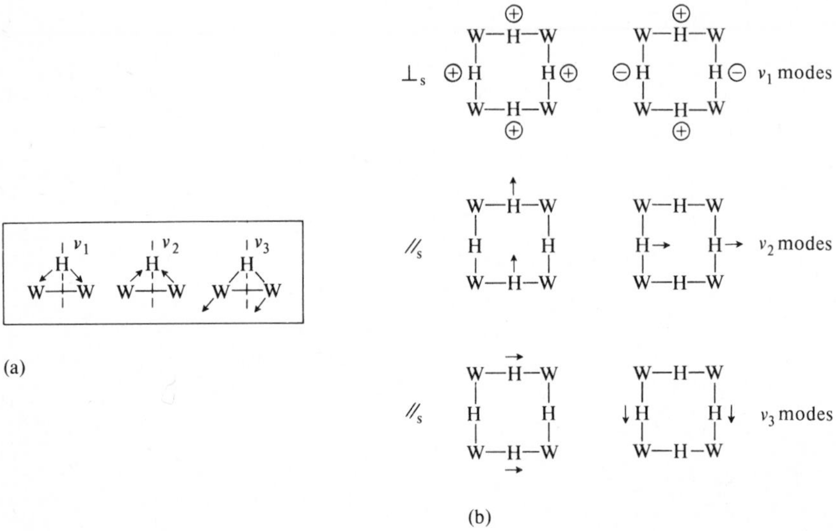

Fig. 7.12. Hydrogenic lattice modes of hydrogen adsorbed on tungsten (100). ν_1 is at 130 meV and is strong. ν_2 is at 80 meV and ν_3 is at 160 meV. Both ν_2 and ν_3 are weak 'forbidden' absorptions. (a) Side view; (b) top view. (From Willis, Ho, and Plummer 1979.)

observations away from the specular angle weak features due to two other 'forbidden' fundamentals were observed. The various vibrations are also illustrated in Fig. 7.12.

Finally, let us summarize some of the current relative strengths and weaknesses of EELS and IR for the study of adsorbed layers.

sensitivity: EELS greater than IR;

resolution: IR better than EELS;

detection of parallel modes of vibration: not by IR; possible by off-specular EELS;

Single crystals: IR almost entirely devoted to CO; EELS gives good spectra for catalytically interesting systems.

8. General background to heterogeneous catalysis and its application to the reaction of hydrogen isotopes

Introduction

Two main strands can be identified in the way studies of gas–solid heterogeneous catalysis have developed. The first is related to the needs of the chemical industry. The prime concern here has been the development of practical catalysts which are specific to the reaction concerned and which operate efficiently under manufacturing conditions. Of necessity, trial and error procedures play a dominant role in the development process of such catalysts. The important consideration is that the catalyst works; how it works is of secondary concern. This practical aspect of heterogeneous catalysis has been admirably covered by Satterfield (1980) and Gates, Katzer, and Schmit (1979). Schuit

The second aspect of catalytic studies has been the attempt to understand at an atomic level the fundamental nature of the catalytic process. This has required ever-improving definition of the catalyst surface and the use of all available physical and chemical techniques for identifying the processes at the gas–solid interface. Naturally enough, the methods for studying systems in which adsorption is the dominant process are often appropriate also for catalytic systems. As we noted in Chapter 2, surface reactions leading to the production of new molecules can be seen as but one of several possible eventual outcomes of the collision of a gas molecule with the surface. Thus all of the techniques discussed in Chapters 5, 6, and 7 have been used to study catalysis and we shall consider some examples of their application in the next chapter.

One difficulty in discussing the mechanisms of heterogeneous catalysis arises from the difference of many orders of magnitude between the pressures at which industrial processes are carried out, upwards of 1 atm, and those at which fundamental studies are done, usually less than 10^{-9} atm. With a factor of 10^{10} between the characteristic pressures, it is little wonder that there is not much overlap between the experimental methods appropriate to each region. One particular point of difficulty arises from the widespread use in fundamental catalytic studies of techniques, some described in previous chapters, which require the passage of charged particles. Low working pressures are then essential to avoid scattering. It is thus not possible to monitor the surface continuously by one of these methods during an industrial catalytic reaction. Furthermore, there is little resemblance between the carefully cleaned and well-characterized surfaces reacting with pure gases, characteristic of laboratory work, and the conditions obtaining in an industrial process. However, a significant advance towards bridging the gap between high pressure and low pressure studies has come from the development of apparatus in which the catalytic activity at practical pressures of previously fully characterized single-crystal surfaces is measured. Subsequent re-evacuation of the reaction chamber allows the surface to be re-examined under the original conditions to determine whether structural changes or other modification to the surface occurs during reaction. All of this is achieved without exposing the surface to the atmosphere. Accounts of some of these elegant experiments have been given by Somorjai (1977, 1979).

Preparation of catalysts

There are three main types of metal surface whose catalytic activity has been investigated. The first is a bulk sample, usually in the form of a single-crystal surface or a fine wire. Preparation and cleaning of these specimens follows exactly the methods used for adsorption studies. It would, indeed, be usual to record the adsorptive as well as the catalytic properties of such a sample. The second type of surface is an evaporated metal film. This is again prepared in the same way as for adsorption studies. The third, which is by far the most important type of catalyst for industrial processes, consists of small particles of the metal supported on a finely divided oxide powder (often Al_2O_3, SiO_2, or MgO). The support is not necessarily inert and when it participates in the catalytic process, the catalyst is said to be 'bifunctional'.

Supported metal catalysts are usually prepared by one of two main procedures. The first involves impregnating the oxide powder with a solution of a salt of the metal. If a bimetallic catalyst, i.e. one with two metals on the surface, is required, a mixture of salts is used. The concentration of the metal salt(s) determines the 'loading' of the oxide surface with the catalytic metal. Usually a loading of about 1 wt. per cent is chosen for expensive catalysts such as platinum. This ensures that only a small proportion of the oxide is covered by metal, and allows very small metal particles to form on the surface. Small particles are desirable since they maximize the fraction of the metal exposed as surface atoms. The impregnated oxide is then dried and the salt is reduced to the metal. The precise method of reduction varies from catalyst to catalyst and is a crucial factor in determining the activity of the resulting sample. The second main method of preparation may be preferred if a high rather than a low loading is required, as for example in base metal catalysts. It involves the precipitation of the metal, usually as the hydroxide, onto the surface of the support. Thus in a typical process an aqueous solution of nickel nitrate is stirred with alumina and ammonium hydroxide is added to the slurry; the nickel hydroxide then precipitates onto the alumina. Following filtration and drying metallic nickel is formed by reduction in hydrogen at 500–600 °C.

General description of heterogeneous catalysis

The familiar long-established definition of a catalyst as a 'substance which increases the rate of a chemical reaction without itself being consumed' has not been improved upon and we shall use it. The question then arises as to the mechanism by which a catalyst works. At the moment no general *ab initio* theory of catalysis is in sight and at the molecular level we shall be limited to correlating catalytic properties with other properties of metals. Two of these properties which have attracted interest are (1) the electronic and (2) the geometric properties of the metal.

Most catalytically active metals are in the transition group so that it is natural to associate their activity with d electrons. Attempts have been made to assess the importance of metallic d-band structure to catalytic activity. The method chosen was to change the electronic occupancy of the d band by alloying it with a metal of similar size. In this way it was hoped to avoid significant changes in surface geometry, whilst altering the electronic properties. However, as we have seen, the surfaces of alloys are now known to differ significantly from the bulk composition, so that the role of this 'electronic factor' in catalysis is uncertain. The influence of surface geometry has been investigated both by using metals in various states of dispersion (i.e. of crystallite size) and by looking at the different reactivities of different single-crystal surfaces. The results show that, whilst each reaction–catalyst system needs to be considered individually, some broad classes of behaviour can be discerned. We shall consider these later. One other aspect of catalytic activity that has been of interest is the role of surface imperfections in catalysis. It is difficult to put this effect on a quantitative footing, but the experiments to be described in the next chapter represent a significant advance.

The lack of a comprehensive electronic theory of catalysis has parallels in conventional reaction kinetics. It is only in the area of reactive molecular beam studies that potential energy curves calculated theoretically are available for comparison with experimental data. Elsewhere, the language of transition state theory is frequently used to discuss reaction rates. We shall therefore enquire how this theory can be applied to catalytic reactions.

Transition state theory of heterogeneous reactions

Our aim is to compare, in order-of-magnitude terms, the rate at which a gas phase bimolecular reaction proceeds in the absence of the catalyst, that is to say the ordinary homogeneous reaction rate constant, with the rate constant for the reaction taking place on the catalyst. The general form of the reaction potential energy curve for the homogeneous reaction is as illustrated in Fig. 8.1. The highest point on the curve is the transition state. The analogous curve for the surface reaction is also shown in Fig. 8.1. Some special features of the catalytic curve should be noticed.

1. An activation energy barrier to adsorption has been included. As we have noticed, this barrier may well be zero.

2. The true activation energy for the surface reaction is from the valley representing the energy of the adsorbed reactants to the surface transition state. However, the temperature dependence of the reaction rate does not necessarily depend on this energy alone, as we shall see later in this chapter.

3. Excitation to the surface transition state is assumed to be the

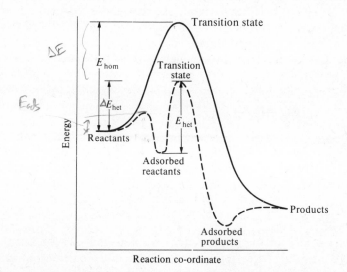

Fig. 8.1. Schematic representation of the transition state theory of a homogeneous reaction (————) and a heterogeneous reaction (– – –).

rate-determining step. One consequence of this assumption is a direct proportionality between surface concentration and reaction rate; another is that adsorption and desorption are not the rate-limiting processes.

It is also assumed that the overall rate of the reaction is directly proportional to the amount of catalyst added to the system.

For a homogeneous bimolecular reaction, the rate of reaction between reactants A and B is given by

$$R_{hom} = c_A c_B A \exp\left(\frac{-E_{hom}}{RT}\right)$$

where c is the concentration. In terms of transition state theory

$$R_{hom} = c_A c_B \frac{kT}{h} \frac{(q_{\neq})_{hom}}{q_A q_B} \exp\left(\frac{-E_{hom}}{RT}\right)$$

The partition functions q_A, q_B, and $(q_{\neq})_{hom}$ relate respectively to the reactants and the transition state. In the case of $(q_{\neq})_{hom}$ the theory has factorized out one degree of freedom, namely motion along the reaction co-ordinate, in order to obtain the pre-exponential factor kT/h. The partition function $q_{\neq}$ is thus related to the full partition function $q^{\neq}$ of the transition state by

$$q^{\neq} = \frac{kT}{h} q_{\neq}.$$

For the heterogeneous reaction we make the simplifying assumption that the reaction is between surface species whose concentration is directly proportional to the gas phase concentration, and that the reaction takes place at adjacent pairs of surface sites S_2 whose concentration is c_{S_2}. (This is the Langmuir–Hinshelwood mechanism—for which see later). Then

$$R_{het} = c_A c_B c_{S_2} \frac{kT}{h} \frac{(q_{\neq})_{het}}{q_A q_B} \exp\left(\frac{-E_{het}}{RT}\right).$$

Therefore the ratio of the states is

$$\frac{R_{het}}{R_{hom}} = c_{S_2} \frac{(q_{\neq})_{het}}{(q_{\neq})_{hom}} \exp\left(\frac{\Delta E}{RT}\right)$$

where $\Delta E = E_{hom} - E_{het}$.

To obtain an idea of the relative rates we note that (1) for a typical surface there are $\sim 10^{15}$ sites cm^{-2} and (2) the modified partition function in the gas phase includes a contribution from translational motion, whereas this motion is lost on adsorption. Since the partition function for translation is far greater than for other sorts of molecular motion, the partition function ratio is approximated by

$$\frac{(q_{\neq})_{het}}{(q_{\neq})_{hom}} \approx \frac{1}{q_{translation}}.$$

Translational partition functions may take a considerable range of values, but a typical number would be 10^{27}. Thus, and very roughly, for a catalyst of area $1\ cm^2$

$$\frac{R_{het}}{R_{hom}} \approx \frac{10^{15}}{10^{27}} \exp\left(\frac{\Delta E}{RT}\right).$$

Therefore, for the heterogeneous process to compete with the homogeneous, a substantial lowering of the activation energy is required; in this example a reduction of about $67\ kJ\ mol^{-1}$ at 300 K is needed to make the rates equal. Finally, we should note the simplifications and assumptions in this derivation, particularly perhaps the limitation of surface area to $1\ cm^2$. Practical catalysts can have areas several orders of magnitude greater than this value, so that overall the reduction in activation energy of a successful catalyst may be considerably less than the figure derive above.

Specificity and selectivity in catalysis

The role of a catalyst is to increase the rate of attainment of equilibrium; if an added substance alters the position of the equilibrium it is no catalyst. However, for reactants of even modest complexity, more than one set of products may be thermodynamically possible. Which particular reaction pathway is favoured may depend upon the choice of catalyst, and there is no presumption that the thermodynamically most stable set of products will predominate. There has been discussion of two related questions arising from these considerations. The first question is how the rate of a particular reaction can be expected to vary if the catalyst is modified. Such modification can be achieved by the major step of moving from one metal to another or, more subtly, by alloying with another metal. The pattern of behaviour resulting from systematic investigation of this question has been referred to as the 'catalytic specificity' (Boudart 1979). The second question relates to the way in which the relative rates of different surface reactions change when the catalyst is modified. In this case modification has also included changing the surface of a particular metal, e.g. from one crystal plane to another, or by altering the degree of dispersion (i.e. the fraction of metal atoms at the surface) of a supported metal catalyst.

The relative rates of heterogeneous reactions taking place in parallel on a catalyst can be used to define how 'selective' the catalyst is. Thus, if several catalytic reactions are occurring, the selectivity S for a chosen product i is defined by

$$S = \frac{\xi_i}{\sum_i \xi_i}$$

where ξ is the rate of reaction. Alternatively, and particularly when two reaction pathways i and j predominate, the selectivity can be defined by the ratio of the two rates, i.e.

$$S = \xi_i/\xi_j.$$

It has been suggested that a comprehensive review of all catalytic reactions studied is unlikely to shed much light on the factors which influence selectivity and specificity, since the range of mechanisms and conditions is too wide. Accordingly we shall limit our discussion to certain aspects of the reactivity of the group VIII metals (see Boudart 1979) for which some patterns of behaviour are to be discerned. For this purpose we shall take a group of reactions involving hydrogen and one other molecule and consider the results of experiments in which the effects of surface structure, change of metal and alloying on the catalytic activity were observed. The change of surface structure was brought about either by altering the particle sizes of supported metal catalysts or by varying the crystallographic planes exposed by single-crystal surfaces. Alloying was carried out with group Ib metals (usually copper or gold) which are catalytically inert. Stepped as well as low index planes have been investigated for some of the metals. The reactions chosen are

(1) $H_2 + D_2 \rightarrow 2HD$ equilibration

(2) $C_2H_4 + H_2 \rightarrow C_2H_6$ hydrogenation

(3) cyclo-$C_3H_6 + H_2 \rightarrow C_3H_8$ hydrogenation

(4) $C_6H_6 + 3H_2 \rightleftharpoons C_6H_{12}$ hydrogenation/dehydrogenation

(5) $C_2H_6 + H_2 \rightarrow 2CH_4$ hydrogenolysis

(6) $N_2 + 3H_2 \rightleftharpoons 2NH_3$ ammonia synthesis/decomposition.

The results indicated much the same pattern of behaviour for reactions (1)–(4) on the various metals. As Table 8.1 shows, there was hardly any effect of surface structure (less than 10-fold), a small effect of alloying (10–100-fold), and a moderate effect of changing from one metal to another. Perhaps the most surprising result here is the absence of any sensitivity of benzene hydrogenation to the surface structure. Intuitively, a

Table 8.1
Effects of metal structure, nature of the metal, and alloying on activity for catalytic reactions

Reaction	Effect of structure	Specificity	Effect of alloying
$H_2 + D_2 \rightarrow 2HD$	VS	M	S
$C_2H_4 + H_2 \rightarrow C_2H_6$	VS	M	S
cyclo-$C_3H_6 + H_2 \rightarrow C_3H_8$	VS	M	S
$C_6H_6 + 3H_2 \rightleftarrows C_6H_{12}$	VS	M	S
$C_2H_6 + H_2 \rightarrow 2CH_4$	S	VL	L
$N_2 + 3H_2 \rightleftarrows 2NH_3$	S	L	?

Scale: 1   10 10^2 10^4 10^6 UP
          VS   S   M   L   VL

From Boudart 1979.

significant dependence on geometry would be expected for the reaction of as complicated a molecule as benzene. This class of reaction, which is independent of surface structure has been called 'facile' or 'structure insensitive'.

In contrast with reactions (1)–(4), reactions (5) and (6) are markedly sensitive to all three variables. Thus, for example, hydrogenolysis of ethane depends somewhat on the size of particles of the metal (10–100-fold) but enormously on the metal ($\sim 10^8$-fold as between Os and Pt). Alloying also has a considerable effect, 6 per cent of copper added to nickel reduces the activity a thousand-fold. Ammonia synthesis and decomposition also varies greatly from metal to metal, and there is some effect of particle size. Reactions with these properties have been called 'demanding' or 'structure sensitive'.

Catalytic activity and the strength of chemisorption

In view of the inextricable link between heterogeneous catalysis and chemisorption it is worthwhile considering how catalytic activity might be expected to change as the strength of adsorption increases. Qualitatively we can see that two limiting situations may arise. If the chemisorption is 'weak' the surface coverage will be sparse and correspondingly the catalytic activity will be low. Since an increase in a low heat of adsorption increases the uptake, activity will increase in this region with strength of adsorption. At the other extreme of 'strong' adsorption, the surface will be essentially fully covered. However, the surface species may then be so stable that it is difficult to decompose it. Again catalytic activity will be feeble. In this case a decrease in strength of adsorption will lead to increased catalytic activity. The overall result is thus that we should expect the activity curve to pass through a maximum as the strength of adsorption passes from 'weak' to 'strong'. This peak-shaped plot is known rather picturesquely as a Balandin 'volcano plot'. However, when it comes to selecting the experimental data to represent 'catalytic activity' the choice is by no means self-evident. The velocity constant k for a catalytic reaction is an obvious possibility. However, k includes the Arrhenius parameters A and E_{act}, both of which may vary when the catalyst is changed. If they vary in the same direction, the overall change in k at a particular temperature is less than it would be if only the activation energy were varying. This effect is by no means uncommon and is referred to as the 'compensation effect' since the change in A partially compensates for the change in E_{act}.

Accepting, therefore, that there is an arbitrary element in the choice of criterion for defining catalytic activity, let us consider a reaction which elegantly exemplifies the volcano plot. This reaction is the decomposition of formic acid vapour on a series of transition metals (Rootsaert and Sachtler 1960). The interaction between formic acid and the metal surfaces is sufficiently strong that a surface compound is formed, namely the metal formate. A suitable measure of the strength of interaction at the surface is then the heat of formation of the metal formates. For this reaction, the criterion chosen for defining catalytic activity was the metal temperature T_r, at which the rate of decomposition of formic acid reached a chosen value. Thus the more catalytically active was the metal the lower was T_r. The resulting plot is shown in Fig. 8.2.

We can give a more quantitative meaning to a curve such as that shown in Fig. 8.2 by taking together our treatment of the Langmuir isotherm in Chapter 1 and the potential energy profile illustrated in Fig. 8.1. We start with the Langmuir isotherm (eqn (1.6)):

$$\theta = \frac{bp}{1 + bp}$$

where

$$b^{-1} = (2\pi mkT)^{1/2} \frac{k_d}{s^*} \exp\left(\frac{\Delta H_{ad}}{RT}\right)$$

or

$$b^{-1} = c \exp\left(\frac{\Delta H_{ad}}{RT}\right).$$

(s^* and k_d are assumed not to vary significantly in the range of θ under discussion.) When we are dealing with just one surface species, we can

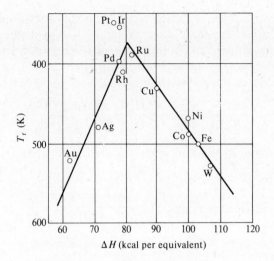

Fig. 8.2. An example of a 'volcano plot', the decomposition of formic acid. T_r is the temperature at which the rate of decomposition reaches a specified value. (From Rootsaert and Sachler 1960.)

write for the rate of reaction on the surface

$$R_{\text{het}} = k_{\text{het}}\theta.$$

Let us take first the limiting case of a numerically low heat of adsorption; b is then small (remember that ΔH_{ad} is negative) and $bp \ll 1$, so that

$$\theta = bp = \frac{p}{c}\exp\left(\frac{-\Delta H_{\text{ad}}}{RT}\right)$$

and

$$R_{\text{het}} = k_{\text{het}}\frac{p}{c}\exp\left(\frac{-\Delta H_{\text{ad}}}{RT}\right).$$

If we now write k_{het} in terms of the Arrhenius equation, the relevant activation energy is E_{het} in Fig. 8.2. Thus

$$R_{\text{het}} = A_{\text{het}}\frac{p}{c}\exp\left(\frac{-\Delta H_{\text{ad}}}{RT} - \frac{E_{\text{het}}}{RT}\right). \tag{8.1}$$

Now let us consider as a model two catalytic reactions characterized by the same activation energy for the surface reaction but with different low enthalpies of adsorption $(\Delta H_{\text{ad}})_1$ and $(\Delta H_{\text{ad}})_2$. Then these systems have equal rates of reaction at the corresponding values of the characteristic temperature, which we shall call $(T_r)_1$ and $(T_r)_2$. The temperature dependence of R_{het} is dominated by the exponential term and we shall not need to consider the pre-exponential factors further.

Therefore

$$\frac{(R_{het})_1}{(R_{het})_2} = \frac{\exp\{-(\Delta H_{ad})_1/RT - E_{het}/RT\}}{\exp\{-(\Delta H_{ad})_2/RT - E_{het}/RT\}}.$$

When the rates of the two reactions are equal, which is of course at different temperatures, the exponential terms are equal. So we can write

$$\frac{(\Delta H_{ad})_1 + E_{het}}{(T_r)_1} = \frac{(\Delta H_{ad})_2 + E_{het}}{(T_r)_2}.$$

Now let

$$(\Delta H_{ad})_2 = (\Delta H_{ad})_1 + \delta\Delta H$$

where $|\delta\Delta H|$ is the increase in the magnitude of the heat of adsorption on passing from catalyst 1 to catalyst 2 (since the enthalpy of adsorption is negative, it follows that $\delta\Delta H$ is also negative for an increased interaction between catalyst and gas). Substituting and rearranging we find that

$$(\Delta H_{ad})_1 + E_{het} = \frac{(T_r)_1 \delta\Delta H}{(T_r)_2 - (T_r)_1}$$

$$= \frac{(T_r)_1 \delta\Delta H}{\Delta T} \tag{8.2}$$

where

$$\Delta T = (T_r)_2 - (T_r)_1.$$

Equation (8.2) gives us a quantitative measure of the way in which the characteristic temperature should change with the enthalpy of adsorption. In order to relate this equation to the volcano curve, we note first that we are dealing with the left-hand slope of the volcano where adsorption is weak. Turning to eqn (8.2), we further note that the left-hand side is always positive since the activation energy for the surface reaction is numerically larger than the enthalpy of adsorption. Thus, on the right-hand side of eqn (8.2) we can see that an increase in the gas–solid interaction, a negative value of $\delta\Delta H$, requires that ΔT is negative, i.e. that $(T_r)_2 < (T_r)_1$. This is in accord with the experimental observations. Furthermore, eqn (8.2) predicts that there should be a linear relationship between $\delta\Delta H$ and ΔT. This again agrees very satisfactorily with Fig. 8.2.

Now let us turn to the far right-hand slope of the volcano where there is strong adsorption, and ΔH_{ad} is numerically large. In this case using the Langmuir isotherm and a numerically large value of ΔH_{ad} yields $bP \gg 1$ and $\theta \rightarrow 1$. This is the conclusion that for a strongly adsorbed species the surface is fully covered. Thus the surface rate is given by

$$R_{het} = k_{het}$$

and is independent of pressure, i.e. a zero-order reaction. A modest change in ΔH_{ad} would still leave $bP \gg 1$ and thus at first sight we would expect the rate to be unchanged, which is contrary to observation.

In choosing a model for the reaction to account for the observed change of rate we make the postulate that changes in binding energy affect the bottom of the potential energy curve for the surface reaction, i.e. the adsorbed molecule, much more than the transition state. There is some parallel for such an assumption in the discussion of the kinetic isotope effect in solution kinetics. The difference between the velocity constants for hydrogen and deuterium transfer reactions is attributed to a larger effect of isotopic substitution on the initial reactant than on the transition state (see Pilling 1975).

The model implies that the heterogeneous activation energy can be looked on as having two contributions. These are (1) the enthalpy of adsorption and (2) the energy difference between the reactants and the transition state, marked ΔE_{het} on Fig. 8.1. It is this latter quantity that is assumed to be insensitive to the strength of adsorption. Then

$$E_{het} = -\Delta H_{ad} + \Delta E_{het}$$

and

$$R_{het} = k_{het} = A_{het} \exp\left(\frac{-\Delta H_{ad} + \Delta E_{het}}{RT}\right)$$

Thus as the enthalpy of adsorption becomes less negative, the total activation energy decreases. The reaction therefore proceeds faster in the region of strong adsorption as the strength of the adsorption decreases. There is again a linear relationship between T_r and ΔH_{ad}, provided that ΔE_{het} is essentially constant.

It thus appears that, despite the simplifications in the models chosen, a satisfactory interpretation of the general shape of the volcano curve can be given. It should not be expected, however, that such curves would always have linear flanks, though qualitatively similar features are to be expected in other examples.

In order to proceed further with a general discussion of the way individual catalytic systems behave, especially of how they are affected by changes in pressure and temperature, it is necessary to adopt a model for the surface reaction. Two such models have been of prime importance and we shall discuss them in turn. They are the Langmuir–Hinshelwood and Eley–Rideal mechanisms.

The Langmuir–Hinshelwood mechanism

The main assumptions made in this mechanism can be summarized as follows:
   1. the surface reaction is the rate-determining step;
   2. the Langmuir isotherm can be applied to describe the equilibrium between gas phase and adsorbed reactants;
   3. the adsorbed reactants compete for surface sites;
   4. in the case of bimolecular reactions the reaction is between two adsorbed species.

Let us now investigate how this model predicts the pressure and temperature dependence of the rates of some unimolecular and bimolecular surface reactions.

1. Unimolecular surface reactions

Decomposition of a single adsorbed species

The reaction is then

$$A_{ads} \rightarrow products$$

$$rate = k\theta_A.$$

From the Langmuir isotherm

$$\theta_A = \frac{b_A p_A}{1 + b_A p_A}$$

Therefore

$$rate = \frac{k_{het} b_A p_A}{1 + b_A p_A}.$$

Consider the following two limiting cases.
   (a) $b_A p_A \ll 1$, rate $= k_{het} b_A p_A$, and the reaction is first order in gas pressure.
   (b) $b_A p_A \gg 1$, rate $= k_{het}$, and the reaction is zero order in gas pressure. For a strongly adsorbed species, the Langmuir constant b is large and the inequality $b_A p_A \gg 1$ may obtain for all accessible values of p_A and the observed reaction order is always zero. For an intermediate interaction, the reaction order varies with pressure.

The temperature dependence of the velocity constants can be derived by substituting for k using the Arrhenius equation. Thus for case (a), the experimentally observed rate constant k_{exp} is

$$k_{exp} = k_{het} b_A$$

and

$$\frac{d(\ln k_{exp})}{dT} = \frac{d(\ln k_{het})}{dT} + \frac{d(\ln b_A)}{dT}$$

$$= \frac{E_{het} + \Delta H_{ad}}{RT^2}.$$

Thus the slope of an Arrhenius plot (a graph of $\ln k_{exp}$ versus $1/T$) yields an experimental activation energy E_{exp}, which is smaller than the activation energy for the surface reaction (ΔH_{ad} is negative).

For case (b), $k_{exp} = k_{het}$ and the Arrhenius plot does give the surface activation energy.

Decomposition of a single species (*product also adsorbed*)

The reaction scheme is

$$A_{ad} \rightarrow B_{ad} \rightarrow B_{gas}.$$

The rate of reaction now depends upon how well A competes with the product B for adsorption sites. We have already seen that the coverage of A in a mixed adsorption is given by (see Chapter 1)

$$\theta_A = \frac{b_A p_A}{1 + b_B p_B + b_A p_A}$$

so the rate is

$$R = \frac{k_{het} b_A p_A}{1 + b_B p_B + b_A p_A}.$$

The inhibiting effect of adsorption of the product B is illustrated by two limiting cases:

1. at low pressures of A, where $b_A p_A \ll 1$, then

$$R = \frac{k_{het} b_A p_A}{1 + b_B p_B};$$

2. when the product is strongly adsorbed, so that $b_B p_B \gg 1$ and $b_B p_B \gg b_A p_A$, then

$$R = k_{het} \frac{b_A p_A}{b_B p_B}.$$

2. Bimolecular surface reactions

For bimolecular reactions the competition for surface sites is between the two reactants and the products. A considerable complexity is thus introduced into the reaction schemes. We shall consider some, but not all, of the possible consequences.

Reactants adsorbed; products not adsorbed

The reaction is

$$A_{ad} + B_{ad} \rightarrow \text{products}$$

and the rate is given by

$$R = k_{het} \theta_A \theta_B.$$

Substituting the Langmuir isotherm for θ_A and θ_B yields

$$R = \frac{k_{het} b_A p_A b_B p_B}{(1 + b_A p_A + b_B p_B)^2}. \tag{8.3}$$

Equation 8.3 has some interesting features.

1. If b_A and b_B are of comparable order of magnitude, i.e. neither the A term nor the B term is of overwhelming importance in the denominator, the rate of reaction goes through a maximum as p_B is increased whilst p_A is kept constant.

2. If both A and B are weakly adsorbed, then

$$b_A p_A \ll 1 \quad \text{and} \quad b_B p_B \ll 1.$$

Thus

$$R = k_{het} b_A b_B p_A p_B$$

and the reaction is second order in gas pressure.

3. If one reactant, say A, is weakly adsorbed,

$$b_A p_A \ll b_B p_B + 1$$

and

$$R = \frac{k_{het} b_A b_B p_A p_B}{(1 + b_B p_B)^2}. \tag{8.4}$$

Therefore the reaction is first order in the pressure of A and goes through a maximum rate as the pressure of B is increased at constant p_A.

If as well as A being weakly adsorbed, B is strongly adsorbed, eqn (8.4) can be simplified by noting that

$$b_B p_B \gg 1,$$

so that

$$R = k_{het} \frac{b_A}{b_B} \frac{p_A}{p_B}.$$

The reaction is thus inhibited by an increase in p_B and the experimental rate constant is

$$k_{exp} = k_{het} \frac{b_A}{b_B}.$$

The experimental activation energy is then

$$E_{exp} = E_{het} + \Delta H_A - \Delta H_B.$$

Since B is strongly adsorbed and A is weakly adsorbed

$$|\Delta H_B| > |\Delta H_A|$$

and $E_{exp} > E_{het}$. The strong adsorption of one reactant thus tends to slow down the reaction.

Both reactants and products adsorbed

The reaction scheme is

$$A_{ad} \rightarrow B_{ad} \rightarrow C_{ad} \rightarrow C_{gas}$$

Then

$$\theta_A = \frac{b_A p_A}{1 + b_A p_A + b_B p_B + b_C p_C}$$

and θ_B is given by an analogous expression.

Then the rate of reaction is

$$R = k_{het} \theta_A \theta_B$$

or

$$R = \frac{k_{het} b_A p_A b_B p_B}{(1 + b_A p_A + b_B p_B + b_C p_C)^2}.$$

The most interesting approximation of this equation is for strongly adsorbed products, i.e. when

$$b_C p_C \gg 1 + b_A p_A + b_B p_B$$

so that

$$R = \frac{k_{het} b_A b_B p_A p_B}{b_C^2 p_C^2}.$$

The reaction is then strongly inhibited by the products.

3. Reactions with a maximum rate

One of the features of catalytic reactions which is seldom encountered in ordinary reaction kinetics is that the Arrhenius plot of $\ln k$ versus $1/T$ may pass through a maximum as the temperature is increased. The fundamental reason for this is the interplay between the normal tendency for the surface reaction, characterized by the activation energy E_{het}, to go faster and the tendency for the surface coverage to decrease at higher temperatures. Let us now consider as an example a bimolecular surface reaction and see how the maximum might occur.

The rate of the surface reaction is given in eqn (8.3) as

$$R = k_{het}\theta_A\theta_B$$

$$= k_{het}\frac{b_A p_A b_B p_B}{(1 + b_A p_A + b_B p_B)^2}.$$

Suppose that B is much more strongly adsorbed than A. Then

$$b_B p_b \gg b_A p_A$$

and

$$R = \frac{k_{het} b_A b_B p_A p_B}{(1 + b_B p_B)^2}.$$

Now let us choose conditions of temperature and pressure such that $b_B p_B \gg 1$ (i.e. high p and low T); then

$$R = \frac{k_{het} b_A p_A}{b_B p_B}.$$

Since k_{het} and b depend exponentially on $-E_{het}$ and ΔH_{ads} respectively, the slope of an Arrhenius plot has a slope determined by $E_{het} + (\Delta H_{ad})_A - (\Delta H_{ad})_B$. Bearing in mind that the ΔH_{ads} is negative and that we have chosen B to be more strongly adsorbed than A, it follows that $(\Delta H_{ad})_A - (\Delta H_{ad})_B > 0$. Overall, therefore, the experimental activation energy $E_{het} + (\Delta H_{ad})_A - (\Delta H_{ad})_B$ is positive and the Arrhenius plot has a normal slope; k increases with increase in temperature.

Next let us alter the conditions of the experiment so that $b_B p_B \ll 1$ (i.e. low p and high T). Thus

$$R = k_{het} b_A b_B p_A p_B.$$

The slope of the Arrhenius plot is now determined by the energy terms

$E_{het} + (\Delta H_{ad})_A + (\Delta H_{ad})_B$. Remembering that we have postulated that B is strongly adsorbed, there is no reason in principle why the numerical magnitudes of the heats of adsorption should not exceed E_{het}. If this happens $E_{het} + (\Delta H_{ad})_A + (\Delta H_{ad})_B$ becomes negative, that is to say the reaction goes slower at higher temperatures.

Overall, therefore, a plot of $\ln k_{exp}$ versus $1/T$ passes from a positive slope (E_{act} is negative) to a negative slope, as $1/T$ increases, and $\ln k_{exp}$ passes through a maximum.

Throughout this discussion it has been assumed that the surface reaction is the rate-determining step. However, this is not always so. Strong adsorption may cause the vacating of surface sites to become rate determining. When this happens the rate of the reaction increases with increase in temperature as a consequence of the decrease in surface coverage. Eventually, however, when the temperature is high enough, the surface coverage becomes sparse. At this stage the rates of both surface reaction and of desorption are likely to be rapid and neither is then rate determining. Under these circumstances the rate of adsorption may take over as the limiting step. If this happens the high temperature reaction velocity constant is simply the sticking probability. Reactions proceeding via a Langmuir–Hinshelwood mechanism for which this behaviour is quite common are the catalytic equilibrations of a mixture of isotopes. Typical examples are

$$^{15}N_2 + {}^{14}N_2 = 2{}^{15}N^{14}N$$

and

$$^{12}C^{18}O + {}^{13}C^{16}O = {}^{12}C^{16}O + {}^{13}C^{18}O$$

catalysed by transition metals.

It often happens that the reaction first becomes observable when the chemisorbed gas starts to desorb at a temperature known from flash filament experiments. The rate then rises rapidly with an increase in temperature, as θ decreases, until the high temperature region is reached when the rate of the reaction is determined by the reactive sticking probability on the sparsely covered surface. This latter quantity is often insensitive to temperature. A good way of following isotope equilibration reactions and of measuring high temperature sticking probabilities is by mass-spectrometric measurements of the composition of a mixture of isotopes passing over a heated metal filament. The greater is the partial pressure of the product isotopes the faster will be the reaction. Two typical examples are shown in Fig. 8.3.

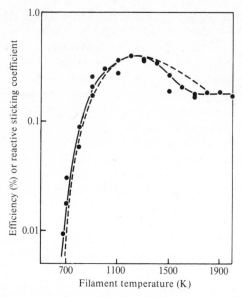

Fig. 8.3. A comparison of two independent measurements of the efficiency of rhenium as a catalyst for the isotope equilibration reaction $^{12}C^{18}O + {}^{13}C^{16}O \rightarrow {}^{12}C^{16}O + {}^{13}C^{18}O$. (From Gasser and Holt 1977.)

The Eley–Rideal mechanism

In contrast to the Langmuir–Hinshelwood mechanism, in which both reacting species are adsorbed, the Eley–Rideal mechanism arises from the possibility that an adsorbed species may react with a gas molecule by a collision mechanism. The reaction scheme is

$$A_{ads} + B_{gas} \rightarrow \text{product.}$$

The rate is then given by

$$\text{rate} = k'_{het}\theta_A p_B \tag{8.5}$$

(k'_{het} is the Eley–Rideal heterogeneous velocity constant).

Applying the Langmuir isotherm to the adsorption of A we find

$$R = k'_{het} \frac{b_A p_A p_B}{1 + b_A p_A}.$$

If now the rate of reaction is plotted as a function of p_A whilst keeping p_B constant, a limiting value is reached at high values of p_A. This result contrasts with the Langmuir–Hinshelwood mechanism which gives rise to a maximum.

The overall rate of reaction is always first order in p_B. As far as p_A is concerned, the order varies in the same way as for a unimolecular reaction, considered above. That is to say there is a variation from first order in p_A when $p_A b_A \ll 1$ to zero order when $p_A b_A \gg 1$. The temperature dependence of the rate, and thus the numerical value of E_{het}, can be deduced in the same way as for the Langmuir–Hinshelwood mechanism.

The transition state associated with the Eley–Rideal mechanism involves the formation of a surface complex comprising the adsorbed species plus the incoming gas molecule. The latter may be physically located either on a weakly held site on the surface adjacent to the adsorbed molecule, or attached to the adsorbed molecule by the transient bonding of a transition state, e.g.

$$M \begin{matrix} \cdots H \cdots \\ \\ \cdots H \cdots \end{matrix} H$$

General considerations in the determination of heterogeneous reaction mechanisms

The measurement of the pressure and temperature dependences of the rates of a heterogeneous reaction are a necessary but not always a sufficient procedure for determining the mechanism. Unfortunately, reactions tend not to fall neatly into one of the limiting cases for which simple mathematical formulations are possible, some of which have been illustrated in the preceding sections. For example, the experimentally accessible range of pressures over which a reaction can be studied may be inadequate to change the surface coverage enough for the dependence of rate on θ to be determined. Changing the temperature can help in this investigation but of course usually has the complicating effect of altering the velocity constant. However, measurements of the dependence of reaction rates on temperature have indeed played a major part in assigning mechanisms to heterogeneous reactions. A further complexity to be borne in mind is the possibility that a heterogeneous reaction may change its mechanism when the temperature is changed.

Other measurements which have played an important role in determining mechanisms are rates of adsorption and desorption. Thus a comparison of the observed catalytic reaction rate with the adsorption or desorption rate under similar conditions is frequently helpful. Such comparisons may show, as we saw for isotope equilibration reactions, that adsorption or desorption, rather than the surface reaction, is rate determining. Alternatively, considerable catalytic activity may be observed under conditions where the rate of molecular interchange between the gas and the surface is known to be slow. Such observations would point to the Eley–Rideal rather than the Langmuir–Hinshelwood mechanism.

The diversity of heterogeneous reactions that have been studied is so great that little further systematic classification is practicable. We shall proceed therefore by way of a choice of examples of catalytic reactions which illustrate the ways in which the techniques described so far have been used to try to establish catalytic mechanisms. We shall discover in so doing how difficult an unambiguous assignment of mechanism can be and that the experimenter in this area needs to be aware that proposed mechanisms may very well need revision in the light of new evidence.

One set of reactions which has attracted interest over a long period of years and played a valuable part in the evolution of theory are those of the hydrogen isotopes. A discussion of these reactions occupies the remainder of this chapter.

Catalytic reactions involving hydrogen isotopes

1. *Ortho*-hydrogen and *para*-hydrogen

Homonuclear diatomic molecules whose nuclei possess nuclear spin can exist in two forms, known as the *ortho* and *para* states. In the case of hydrogen the nuclear spin is $\frac{1}{2}$, and in the hydrogen molecule these nuclear spins can take up one of two alignments which can be depicted as either ↑ or ↓. The result is that the total nuclear spin of the molecule can be either (↑↑) or (↑↓). The spin-parallel state (↑↑) is the *ortho* state and the spin-paired state (↑↓) is the *para* state.

At room temperature hydrogen consists of a mixture of three parts o-H_2 to one part p-H_2. However, when the gas is cooled the equilibrium moves in favour of p-H_2, until at the liquefaction temperature almost pure p-H_2 is expected. The reasons for these observations are to be found in textbooks of statistical thermodynamics (e.g. Gasser and Richards 1974). From the point of view of heterogeneous catalysis there are two significant features of o- and p-H_2. The first is that the reversal of a nuclear spin required to interconvert the two states does not take place readily by collision. Thus, if pure p-H_2 is prepared by liquefying hydrogen and then the sample is returned to room temperature, the gas may remain in the *para* form for a considerable time. The second point is that although there are no significant chemical differences between o- and p-H_2, there is a considerable difference in their thermal conductivities. Since the thermal conductivity of a gas is easy to measure, a convenient method of analysing mixtures of o- and p-H_2 is available. Deuterium, which has a nuclear spin of 1, also has *ortho* and *para* states of which the low temperature state this time is o-D_2.

The establishment of the equilibrium composition can be hastened by the presence of a catalyst. Thus, for room temperature studies the interest is in the *para* to *ortho* conversion, whilst at low temperatures the reverse reaction is studied. Two main mechanisms have been envisaged by which a catalyst can reverse a nuclear spin and thus bring about the interconversion. The first is based on a physical interaction which flips a nuclear spin within a molecule. The second is a chemical process involving bond rupture. The physical mechanism requires physisorption of the H_2 molecule and a strong magnetic interaction between the substrate and the magnetic moment of the adsorbed hydrogen nucleus. The necessary interaction is provided by the high magnetic fields present at the surfaces of paramagnetic substances such as transition metals, their oxides, or activated charcoal. These are indeed effective catalysts. The chemical mechanism proceeds by dissociative adsorption of the H_2 molecule, followed by recombination of the H atoms to give H_2 molecules in the equilibrium proportions. These then desorb. This mechanism requires a mobile chemisorptive equilibrium between the gas and surface phases. We can depict these mechanisms schematically as follows.

Physical process:

$$p\text{-}H_2 + P^* \rightarrow P^* \cdot H_2 \rightarrow P^* + o\text{-}H_2 \quad (P^* \text{ is a paramagnetic centre})$$
$$\uparrow\uparrow \qquad\qquad\qquad \uparrow\downarrow$$

$$(8.6)$$

Chemical process:

$$p\text{-}H_2 + M \rightarrow M \cdot H \rightarrow M + o\text{-}H_2 \quad (M \text{ is the metal surface}).$$
$$\uparrow\uparrow \qquad\qquad \uparrow\downarrow$$

There is also a high temperature (>1000 K) homogeneous equilibration mechanism involving atoms. It is written as

$$H_2 \leftrightarrows H + H$$
$$H + p\text{-}H_2 \rightarrow o\text{-}H_2 + H$$

No contribution is made by this process under catalytic conditions.

A difference in the temperature dependences of the two mechanisms is to be anticipated. The physical mechanism requires no activation energy and would therefore have only a slight temperature dependence. Furthermore, it is expected to proceed readily at low temperatures where it is encouraged by the physical adsorption of hydrogen on the catalyst. On the other hand the chemical process would be expected to have a normal positive temperature coefficient of rate for both the Langmuir–Hinshelwood and Eley–Rideal mechanisms.

2. Hydrogen–deuterium equilibration

Extensive studies of the equilibration reaction

$$H_2 + D_2 \rightarrow 2HD$$

together with the spin-reversal work mentioned above have led to considerable elaboration of the proposals for the mechanism of the reaction. The suggested mechanisms for the reaction are (M is a surface metal atom) as follows.

$$H_2 + D_2 + 4M \longrightarrow \begin{array}{cccc} H & H & D & D \\ | & | & | & | \\ \end{array} \longrightarrow 2HD + 4M.$$
$$\text{—M—M—M—M—}$$

$$(8.7)$$

This chemical fully dissociative mechanism bears the names of Bonhoeffer and Farkas. (It is a special case of the Langmuir–Hinshelwood mechanism.) It requires either that four adjacent adsorption sites are available simultaneously or that the atoms are free to migrate over the surface, and that there is a dynamic equilibrium between gas and surface. It is therefore essentially a 'high temperature' mechanism.

$$M - H_{ad} + D_2 + M \rightarrow MH \cdot MD_2 \rightarrow M - D + HD + M$$
$$M - D_{ad} + H_2 + M \rightarrow MD \cdot MH_2 \rightarrow M - H + HD + M.$$

$$(8.8)$$

This is the Rideal mechanism, in which a molecule is weakly molecularly adsorbed on a site adjacent to a chemisorbed atom of the alternate isotope.

$$M - H + D_2 \longrightarrow M \overset{H}{\underset{D}{\diamondsuit}} D \longrightarrow M - D + HD$$

$$(8.9)$$

$$M - D + H_2 \longrightarrow M \overset{D}{\underset{H}{\diamondsuit}} H \longrightarrow M - H + HD.$$

This is the Eley mechanism which involves a three-centre surface transition state. Experimental discrimination between the Rideal and Eley mechanisms is not practicable and they are usually conflated, as we have seen.

Mechanisms (8.8) and (8.9) were suggested to account for the reactivity of surfaces which were fully covered and from which the desorption rate was negligible. They are thus 'low temperature' pathways to reaction.

In addition, a modification of reaction (8.7) in which the reaction takes place between molecules rather than atoms, and a modification of (8.8) and (8.9) in which the surface migration of atoms is the rate-determining step have been proposed.

We should note that all the foregoing chemical mechanisms are also available for the o-H_2–p-H_2 equilibration reaction.

We shall now consider how the experimental evidence from the study of spin-reversal and isotopic equilibration reactions on a nickel wire has been analysed in terms of the above reaction mechanisms.

The experimental conditions chosen were a gas pressure near 1 Torr and a temperature range of 77–400 K. The chosen temperature could be either that of the system as a whole or that of the nickel wire alone, the walls being at 77 or 273 K. No difference in rates was observed between the two methods. Spin reversal in both hydrogen and deuterium was studied, as well as the $H_2 + D_2$ equilibration reaction. At each particular pressure and temperature the time course of all three reactions was described by the first-order rate equation

$$k = \frac{1}{t} \ln\left(\frac{x_0 - x_{eq}}{x_t - x_{eq}}\right) s^{-1}$$

where x_0 is the fraction of p-H_2, o-D_2, or HD present initially, x_t is the fraction present at any time during the reaction, and x_{eq} is the fraction present at equilibrium. Thus the rate constant k of the reaction was observed from a first-order plot.

The overall rate of reaction per unit area R of filament surface is then

$$R = \frac{nk}{A} \text{ mol cm}^{-2} s^{-1}$$

where n is the number of molecules in the reaction chamber and A is taken as the geometric area of the filament. Arrhenius-type plots of $\ln R$ (corrected to a standard pressure of about 1 Torr) versus $1/T$ then gave the apparent activation energy for the reaction. The overall rate showed a variation with pressure, which could be expressed by

$$R = R_0 p_{H_2}^n$$

The value of n was derived from plots of $\log R$ versus $\log p_{H_2}$.

The results of these experiments are shown as Arrhenius plots for the three processes in Fig. 8.4 in which the temperature range is 77–300 K. Spin reversal in o-D_2 was not followed above 300 K, whilst for p-H_2 the rate was such that a rather approximate value of the activation energy

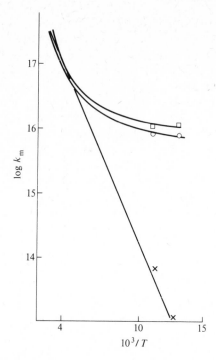

Fig. 8.4. Arrhenius plots for hydrogen isotope reactions on a nickel wire between 77 and 300 K; $\square$, o-H_2–p-H_2; $\bigcirc$, o-D_2–p-D_2; $\times$, $H_2 + D_2$. (From Eley and Norton 1966.)

was derived. However, the $H_2 + D_2$ equilibration could be followed quantitatively up to 400 K and the Arrhenius plot for this reaction between 200 and 400 K is shown in Fig. 8.5.

The resulting apparent activation energies for the various reactions were as shown in Table 8.2. These apparent activation energies are more illuminating than the pressure dependences and we shall discuss them first.

Starting with the low temperature range (77–150 K), the most striking results of this work are observed, namely the non-activated spin-reversal reactions. These reactions proceed even more rapidly than $H_2 + D_2$ equilibration as the temperature is lowered, until at 77 K the ratio of rates is $\sim 10^3$. The conclusion drawn is that spin reversal is predominantly taking place at paramagnetic centres by the non-activated mechanism. Equilibration continues by a chemical mechanism and thus slows down monotonically with decrease in temperature.

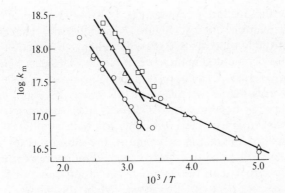

Fig. 8.5. Arrhenius plots for $H_2 + D_2$ equilibration between 200 and 400 K on a nickel wire. The symbols refer to a range of pressures (1.2–3.76 Torr) and to two different wires. (From Eley and Norton 1966.)

In considering the intermediate temperature range(200–300 K), the first point to note is that the $H_2 + D_2$ reaction proceeds with the same activation energy as at the lower temperatures. Presumably, therefore, the reaction mechanism is the same and is thus unchanged over the temperature range 77–300 K. On the other hand, spin reversal is now temperature dependent with an activation energy for o-D_2 which is quite similar to that for equilibration. Although p-H_2 conversion has a some-what lower activation energy (by about 1.2 kcal mol^{-1}), this difference may be due to zero-point energy effects, which are greater for hydrogen than deuterium. The suggestion now is that all three reactions take place by the same mechanism on the surface. Naturally, the physical process for spin reversal continues at the higher temperatures but is overtaken in importance by the chemical process. However, the evidence so far available does not allow an assignment of the chemical mechanism to be made.

Table 8.2

Temperature (K)	Apparent activation energy (kcal mol^{-1})		
	p-H_2	o-D_2	$H_2 + D_2$
77–150	0	0	2.4
200–300	1.5	2.7	2.4
330–400	~5–7	—	7.6

Finally, at the highest temperatures (330–400 K) there is a marked change in activation energy. Figure 8.5 shows the changeover quite clearly and strongly suggests that there is a change in reaction mechanism. The less well documented spin-reversal results suggest a parallel change for this process also. Evidence from previous work indicates that the hydrogen coverage is in the region of 0.5 under the conditions of these higher temperature experiments. At this coverage the Bonhoeffer–Farkas atomic mechanism is expected, probably for both equilibration and spin reversal. If indeed this is the mechanism, the authors (Eley and Norton 1966) considered that the true surface activation energy for equilibration E_{het} would be larger than the apparent activation energy because of the contribution of the heat of adsorption. When this factor is taken into account, a value of $E_{het} = 18$ kcal mol^{-1} is derived.

Having thus assigned a mechanism to the high temperature reaction we can return to the change of mechanism at about 300 K. The proposal made was that this was due to the changeover from the Bonhoeffer–Farkas mechanism (eqn (8.7)) to the Eley–Rideal mechanism (eqns (8.8) and (8.9)). This latter mechanism is thus assigned to the intermediate temperature reaction.

If we turn now to the pressure dependence of the reaction rates, the results are in qualitative rather than quantitative accord with the above proposals for the mechanisms in the various temperature ranges. For the interpretation of the results a modification to the general surface bonding potential energy curve discussed in Chapter 1 and illustrated in Fig. 1.2 was proposed. A further potential energy minimum between the physisorbed state and the strongly chemisorbed state was postulated. This molecular but weakly chemically bonded state had been suggested earlier (Bond 1962) and called type C. The resulting potential energy curve is shown in Fig. 8.6. At the highest temperatures the type C sites are empty and do not contribute to exchange. The reaction is therefore associated with the atomic chemisorbed state which would give rise to a $p^{1/2}$ dependence as is approximately observed at 1 Torr for equilibration (but not for spin reversal). In the range 200–300 K, where the Eley–Rideal mechanism is suggested, this involves chemisorbed atoms plus type C molecules. It has also been suggested that the chemisorption sites would be saturated at these temperatures, leading to zero-order kinetics. However, in practice a range of pressure dependences was found with $n = 0.05$ for o-D_2, $n = 0.22$ for p-H_2, and $n = 0.4$ for $H_2 + D_2$ equilibration. The spin-reversal orders thus approach the expected value, but the equilibration order is anomalously high. This anomaly for equilibration persists down to 77 K, at which n had only fallen to 0.3. On the other hand the expected low fractional order for spin reversal was observed at 77 K where $n = 0.14$ for both p-H_2 and o-D_2.

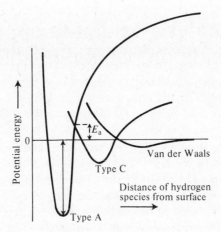

Fig. 8.6. Schematic representation of hydrogen adsorption including the formation of type C adsorption. (From Eley and Norton 1966.)

This example illustrates that even for a simple system the experimentally observed behaviour may not agree in detail with the predictions of the various models. In interpreting the results of any catalytic experiments it is always necessary to consider all the available evidence and then to judge which mechanism most nearly accords with the facts. This cautious attitude will be appropriate in dealing with more complex systems since two significant deviations from the model are to be expected. In the first place it is quite common for heats of adsorption to vary with coverage. As well as the intrinsic heterogeneity associated with evaporated metal films or supported metal catalysts, even single-crystal samples show coverage-dependent heats of adsorption. Another factor, not considered so far, is that the adsorption of one species may modify significantly the bonding of a second species. The heat of adsorption and the activation energy for the surface reaction may thereby become a complex function of the surface coverage.

With these caveats in mind, we turn to a consideration of another simple catalytic reaction, the oxidation of CO.

9. The catalytic oxidation of carbon monoxide on palladium and platinum

Introduction

The catalytic oxidation of CO by palladium and platinum has been intensively studied in the last decade. It is a reaction which is interesting from several points of view. In the first place, the participating molecules contain only two atoms each, so that the surface species formed during the reaction are expected to be relatively simple. There are then grounds for hoping that the mechanism can be understood in some detail. Secondly, the reactions of the gases separately with the metal surfaces lend themselves to study by the physical techniques described in the earlier chapters. A detailed atomic view of the surface layer is thereby made available. Finally, the practical importance of the reaction is immense, in view of the requirement to control the composition of exhaust gases from automobiles in the U.S.A. Platinum or Pt–Pd catalysts, despite their cost, have so far been found to be the cheapest sufficiently durable catalysts for oxidizing hydrocarbons and CO to produce acceptably low levels of these gases in exhaust emissions. Even so, the catalyst is poisoned by lead residues in gases from fuel to which lead alkyls have been added, a formerly common practice for improving performance. Hence unleaded fuel is used in some countries.

As well as palladium and platinum, the other platinum group metals are catalysts for CO oxidation and an excellent comprehensive review of the activity of all the metals is to be found in Engel and Ertl (1979). Only references to work not quoted by Engel and Ertl will be given here.

The plan of the chapter will be to consider first the characterization of the surface layers of CO or oxygen separately on the two metals. In some cases the results and discussion will be considered in detail to exemplify the way in which physical and chemical techniques are used to characterize surface layers. Other results will be quoted without details. We shall then consider mixed layers of oxygen and CO and finally the results and mechanistic interpretation of catalytic experiments.

Adsorbed carbon monoxide

1. Bonding and surface structures

The bonding of CO to palladium or platinum is a typical case of the synergic bonding scheme discussed in Chapter 6 and illustrated in Fig.

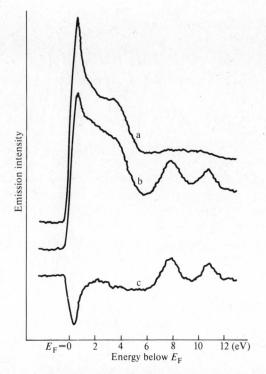

Fig. 9.1. UPS spectra for CO adsorbed on palladium (110): (a) clean surface; (b) with adsorbed CO; (c) difference spectrum (b) − (a). The peak at 8 eV in (c) is due to emission from $5\sigma + 1\pi$ orbitals. The peak at 11 eV is due to emission from the 4σ orbital. (From Küppers, Conrad, Ertl, and Latta 1974.)

6.8. An example of the resulting UV photoelectron spectrum is shown in Fig. 9.1 for CO on palladium (110). As usual the $5\sigma + 1\pi$ orbitals have coalesced to give a single peak near −8 eV (relative to the Fermi level) and the 4σ orbital is near −11 eV. The intense emission starting at the Fermi level and spreading over some 6 eV is associated with the d electrons. Bonding somewhat dampens this emission, producing a dip in the difference spectrum. This dip conceals any contribution to emission from electrons back donated into the $2\pi^*$ orbital which would otherwise be expected in this region. There is no doubt, however, that some back donation is occurring when CO adsorbs on palladium or platinum. The most direct evidence for this comes, as usual, from vibrational spectroscopy. Thus, for example, the reflection–absorption IR spectroscopy band of CO on a platinum (111) oriented ribbon showed a shift from the gas

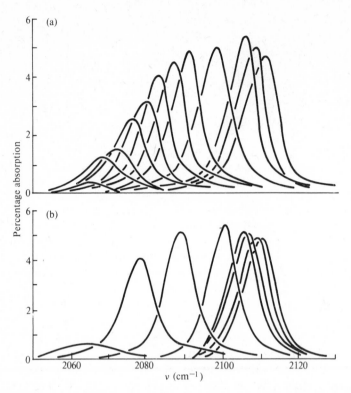

Fig. 9.2. Reflection–absorption IR spectra of CO on platinum (111) at (a) 120 K and (b) 200 K. The band grows in intensity up to $\theta = \frac{1}{3}$; at higher coverages it is constant. The peak frequency increases monotonically with increasing coverage. (From Shigeishi and King 1976.)

phase value of 2143 cm^{-1} to 2065 cm^{-1} at low coverage and moved up in frequency as the coverage increased. The spectra are shown in Fig. 9.2. The frequency range is characteristic of linear-bonded CO.

A satisfactory correlation between IR and LEED observations has been observed for CO adsorbed on palladium (100). In the IR a single band was recorded at all coverages up to $\theta_{CO} = 0.5$. The frequency of this band increased with coverage in the usual way from 1895 cm^{-1} to 1949 cm^{-1} as θ increased from a low value to 0.5. This frequency range points to a bridge-bonded structure. Furthermore, a particular feature of the square array of surface atoms on palladium (100) is that the single set of surface sites associated with just one IR band can be generated only by placing the CO molecules at bridge sites, in conformity with the proposed

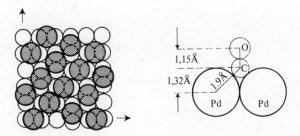

Fig. 9.3. Structural model for CO adsorbed on palladium (100). The CO occupies bridging sites and gives a $(2\sqrt{2} \times \sqrt{2})R45°$ structure at which $[CO]/[Pd] = 0.5$. (From Ertl 1980.)

structure. A dynamical analysis of the LEED intensity data gave the atomic distances and surface structure illustrated in Fig. 9.3. As can be seen these are entirely in accord with the deductions from the IR spectra.

Further, though indirect, evidence about the surface bonding derives from measurements of work function changes following the adsorption of CO. Several single-crystal planes of palladium have been investigated and in each case there is considerable increase in the work function following adsorption. As we saw in Chapter 7, a positive $\Delta\phi$ arises from a transfer of electron density from the metal to the adsorbate. In the case of the Pd—CO surface bond, the resulting dipole moment is about 0.3 debye, though with some variation from plane to plane. This value is large compared with the dipole moment of the gas phase CO molecule, which is only about 0.1 debye. The low gas phase dipole arises from the almost exact balance between the dipole moment of the carbon lone pair $(5\sigma^2$ orbital) and the effect of electron drift from carbon to oxygen in the bonding orbitals as a result of the electronegativity difference between carbon and oxygen. In the surface species Pd–CO, the back donation into the $CO(2\pi)$ antibonding orbitals upsets this balance and leads to the considerable observed increase in dipole moment.

Turning now to the surface structures of CO adsorbed on the low index planes of palladium and platinum, we have to note that there is a special complexity in the case of platinum. This complexity arises from the fact that two of the three commonly encountered low index planes ((100), (110), and (111)), the platinum (100) and the platinum (110), are reconstructed in their stablest form at room temperature. In the case of platinum (100) the reconstructed form is platinum (100) (5×20), although this reverts to the unreconstructed 1×1 pattern following the admission of CO at room temperature. Platinum (110) takes the form platinum (110) (2×1), although the metastable (1×1) can be obtained by heating *in vacuo*. We shall concentrate on the palladium surface structures for which this complication does not occur.

The general feature of the surface structures observed for CO on palladium single-crystal surfaces is the readiness with which one structure evolves into another. At room temperature or below, a series of simple geometric patterns is observed as the CO surface density is increased. Successive patterns are generated by the compression of the existing layer, thus making space available for the extra adsorbed molecules. It should be noted that this process is different from the filling of successive adsorption sites, which was suggested, for example, in the adsorption of CO on polycrystalline tungsten (see Chapter 3). To exemplify the succession of surface structures, we shall consider the hexagonal palladium (111) plane. Models of the surface arrangements are shown in Fig. 9.4. The first structure to develop, shown in Fig. 9.4(a), corresponds to $\theta = \frac{1}{3}$ and is $(\sqrt{3} \times \sqrt{3})R30°$. The assignment of the CO molecules to sites in hollows at which they bond to three palladium atoms is supported by the C—O stretching frequency ($1823\ \text{cm}^{-1}$). Further adsorption compresses the $(\sqrt{3} \times \sqrt{3})R30°$ structure into the $c(4 \times 2)$ illustrated in Fig. 9.4(b) in which the CO molecules have now moved out of the hollows and onto bridge sites. The coverage at this stage is 0.5 and the IR frequency has moved up to $1936\ \text{cm}^{-1}$; this shift is expected and is consistent with bridge sites. Further adsorption can only be achieved by cooling well below room temperature. Hexagonal structures are then formed; an example for which $\theta = 0.66$ is shown in Fig. 9.4(c). This saturated layer has 1×10^{15} CO molecules cm^{-2}.

The hexagonal platinum (111) plane shows similar behaviour. At $\theta = \frac{1}{3}$ the $(\sqrt{3} \times \sqrt{3})R30°$ structure is once again observed. Further adsorption leads to compression into $c(4 \times 2)$, but the final (hexagonal) layer is incoherent.

One particularly interesting feature of these results from the point of

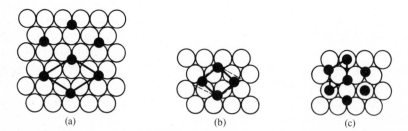

(a)   (b)   (c)

Fig. 9.4. Models for the adsorption of CO on palladium (111): (a) a $(\sqrt{3} \times \sqrt{3})R30°$ structure at $[CO]/[Pd] = \frac{1}{3}$; (b) a $c(4 \times 2)$ structure at $[CO]/[Pd] = 0.5$; (c) a hexagonal structure at $[CO]/[Pd] \approx 0.66$. (From Conrad, Ertl, and Küppers 1978.)

view of catalysis is the readiness with which CO molecules reposition themselves on the surface. The qualitative inference to be drawn from this observation is that the activation energy for surface migration in the chemisorbed state is low. We shall therefore be dealing with a highly mobile CO layer under catalytic conditions. The other point of interest is that so far all the evidence has been consistent with molecular (i.e. non-dissociative) adsorption of CO on palladium and platinum. Indeed there is no evidence to the contrary. The conclusion that adsorption is molecular is consistent with the correlation between heats of adsorption and surface bonding discussed in Chapter 6. The maximum adsorption energy for CO on low index planes of palladium does not exceed 170 kJ mol^{-1} and is thus below that critical value of 260 kJ mol^{-1} at which dissociative adsorption is anticipated.

2. Adsorption and desorption

In this section we shall consider the thermodynamic and kinetic aspects of adsorption and desorption of CO on palladium and platinum. We begin with the heat of adsorption (remember that heat of adsorption = − enthalpy of adsorption). Values for the heat of adsorption have been obtained either from adsorption isobars or by kinetic analysis of thermal desorption spectra. These methods were discussed in Chapters 1 and 3 respectively. An interesting application of changes in work function following adsorption was to the measurement of adsorption isobars of CO on palladium (100). After establishing the direct proportionality between $\Delta\phi$ and θ, at least up to $\theta = 0.8$, work function changes were recorded over a range of temperatures and pressures. The results are shown in Fig. 9.5(a). The variation with coverage of the isosteric enthalpy of adsorption ΔH_{st}, was calculated from these isotherms in the usual way and is shown in Fig. 9.5(b).

The general features which emerge from measurements of heats of adsorption q can be summarized as follows.

1. The initial heat of adsorption varies comparatively little among the low index planes. On palladium the range is from 142 to 167 kJ mol^{-1}, whilst on platinum it is from 109 to 138 kJ mol^{-1}. On stepped surfaces there is some increase in the heat of adsorption in the case of a stepped platinum (111) surface but little effect on stepped palladium (111).

2. The variation of the heat of adsorption with coverage has been most thoroughly investigated for palladium. The effect is fairly small, or zero, depending on the plane, until $\theta \geqslant \frac{1}{3}$. Just above this coverage there may be a stepped reduction in q of up to 30 kJ mol^{-1}. It will be recalled from the LEED discussion that compression of the first overlayer structure on palladium (111), $(\sqrt{3} \times \sqrt{3})R30°$, begins above $\theta = \frac{1}{3}$. There is, therefore, a

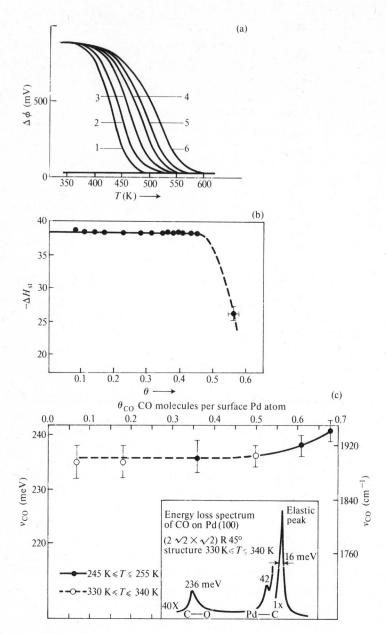

Fig. 9.5. Adsorption of CO on palladium (100): (a) adsorption isobars curve 1, 4×10^{-9} Torr; curve 2, 1.5×10^{-8} Torr; curve 3, 9.8×10^{-8} Torr; curve 4, 3.7×10^{-7} Torr; curve 5, 8.8×10^{-7} Torr; curve 6, 4.4×10^{-6} Torr); (b) coverage dependence of the isosteric enthalpy of adsorption; (c) coverage dependence of ν_{CO} as measured by EELS (the insert shows the complete EEL spectrum). ((a) and (b) from Behm, Christmann, Ertl, and Van Hove 1980; (c) from Behm, *et al.* 1979.)

correlation between the structural and thermodynamic surface properties at this coverage. The correlation continues as θ approaches 0.5, when q drops abruptly and the compression pattern $c(4 \times 2)$ is observed in LEED. At $\theta = 0.5$ the heat of adsorption on palladium (111) has declined to about $90 \, \text{kJ mol}^{-1}$. The need for low temperatures to induce further uptake of CO is now readily explicable, as it is the result of a yet further fall in q as θ is increased.

3. The changes in the IR vibration frequency parallel the changes in q; as the strength of the interaction between CO and the surface, as measured by q, decreases so the C—O stretch frequency increases. An example is the adsorption of CO on palladium (100) for which the experimental results are shown in Fig. 9.5(c). These results accord with the synergic bonding picture, since a decline in surface bond strength, and correspondingly in q, implies reduced occupancy of the antibonding $CO(2\pi)$ orbital and thus increased C—O bonding. None the less, it is worth remembering that repulsive interactions between adsorbed CO molecules, including effects carried through the metal, also contribute IR frequency shifts.

There is one further aspect of the variation of q which is of interest. This is the way in which it might be expected to change with the position of the CO molecule on a clean surface. The difference between peaks and troughs of binding energy would give an indication of the activation energy for migration and thus of surface mobility. Calculations for palladium (100) show rather little variation in energy across the surface and thus predict high mobility—a result in accord with the previous discussion and with experiment. A similar result was obtained for platinum (111).

Having thus considered the thermodynamics of adsorption of CO let us turn to the kinetic aspects: sticking probability curves and desorption kinetics. Sticking probability curves typically show the effect of the formation of a precursor state, as discussed in Chapter 3. An example is the adsorption of CO on platinum (111), which was analysed in terms of the Kisliuk model. The initial sticking probability at 310 K was 0.84 and a satisfactory fit was found for $s_0 = 0.84$ and $K = 0.30$, as illustrated in Fig. 9.6. The initial sticking probabilities vary somewhat in their sensitivity to the plane of the metal and to temperature. Thus on palladium (111), s_0 is 0.96 and is independent of temperature between 300 and 650 K, whilst for platinum s_0 varies from 0.24 on (100) to 1 on (110).

An interesting application of the molecular beam technique has been to the determination of the extent of energy and momentum exchange between the surface of palladium (111) and a beam of CO molecules. In general, specular reflection of molecules leads to a lobe in the reflected beam,

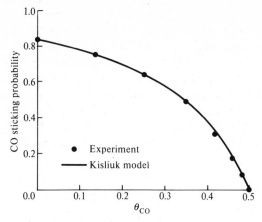

Fig. 9.6. Sticking probability of CO on platinum (111) at 310 K compared with the Kisliuk model ($\theta_{CO} = [CO]/[Pd]$) for $s_0 = 0.84$ and $K = 0.30$ (From Campbell, Ertl, Kuipers, and Segner 1981.)

whereas energy accommodation produces a random desorption pattern which is characterized by a $\cos\theta$ dependence on the reflected angle. Both at 300 K where the [CO]/[Pd] ratio was about 0.5 and at 1020 K where the coverage was very small ($\theta_{CO} < 10^{-6}$) the perfect cosine distribution indicated that complete energy exchange was occurring. The angular distribution is illustrated in Fig. 9.7.

The desorption of the molecularly bonded CO is associated with

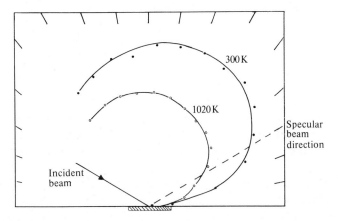

Fig. 9.7. Molecular beam scattering from palladium (111); angular distribution of CO. At 300 K [CO]/[Pd] ≈ 0.5; at 1020 K $\theta_{CO} < 10^{-6}$. (From Engel 1978.)

first-order desorption kinetics. Analysis of desorption curves has provided values for the activation energy for desorption and thus of q. However, these latter calculations often include the assumption that the pre-exponential factor ν_d is $10^{13}\,s^{-1}$. More detailed analyses may show deviations of ν_d from this value, so the adsorption energies thus obtained should be treated with caution. Again, the use of the molecular beam method has proved valuable in providing information about both E_{des} and ν_d for CO desorption from palladium (111). The technique allows the direct determination of the residence time τ of a molecule on the surface. For a first-order desorption

$$\frac{1}{\tau} = k = \nu_d \exp\left(\frac{-E_d}{RT}\right).$$

Thus an Arrhenius-type plot of $\ln \tau$ against $1/T$ should be a straight line of slope E_d/R and intercept ν_d. A plot is shown for temperatures between 550 and 700 K in Fig. 9.8. An excellent straight line is obtained, from which $E_d = 134 \pm 8\,kJ\,mol^{-1}$ and $\nu_d = 10^{14.4 \pm 0.8}\,s^{-1}$ are calculated. In this case E_d agrees with the isosteric heat of adsorption, whilst ν_d is somewhat larger than the $10^{13}\,s^{-1}$ commonly assumed.

This concludes our outline of the adsorption characteristics of CO on palladium and platinum. From the point of view of catalytic oxidation the most significant results seem to be the following:

1. the formation of a monolayer at room temperature in which the ratio of CO to surface metal atoms is roughly $1:2$;
2. the mobility of CO on the metals at room temperature;

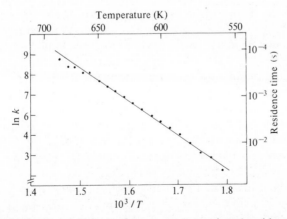

Fig. 9.8. Arrhenius plot of ln (desorption rate constant) and residence time of CO on palladium (111). θ_{CO} is always less than 10^{-2}. (From Engel 1978.)

3. the decline in CO uptake with increased metal temperature to less than 1 per cent of a monolayer at 550 K;

4. an initial sticking probability which is high (0.1–1) at room temperature and remains so at elevated temperatures.

Adsorbed oxygen

1. Bonding and surface structures

The bonding of oxygen to platinum and palladium is more complex than that of CO, since molecular and atomic adsorption can both occur. In addition, incorporation of oxygen atoms into the metal lattice with the formation of a surface oxide is possible. Which of these possibililties is realized depends upon the experimental conditions. Molecular adsorption is a low temperature (<120 K) phenomenon and plays no role in catalysis. At temperatures in the range 150–500 K surface atomic oxygen is the characteristic species. However, the failure of the molecular state at ~100 K to convert to the more thermodynamically stable atomic state indicates that there is an activation energy for this process. On warming a molecular layer to 300 K some oxygen is evolved whilst the remainder converts to the atomic state. Evidence for the nature of these states comes from isotopic mixing experiments; for example on the platinum (111) surface a mixture of $^{16}O_2$ and $^{18}O_2$ in the molecular state does not equilibrate, whereas desorption from the atomic state gives $^{16}O^{18}O$ from $^{16}O_2$ and $^{18}O_2$.

At the highest temperatures (typically > 800 K) a 'surface oxide' which is more difficult to characterize is formed. Indeed it is probable that the transition from chemisorbed layer to 'surface oxide' is gradual. The existence of a form of 'surface' oxygen at elevated temperatures different from the atomic layer is evident from the much higher temperature required to remove it (1250 K compared with > 500 K on platinum (111) or > 700 K on platinum (100)). It is also found that this oxygen is resistant to removal by chemical means such as reduction with hydrogen or CO, whereas the atomic layer is readily removed in this way. That this particularly stable oxygen is below the surface is suggested by ion bombardment experiments (Niehus and Comsa 1981). The scattering of a beam of helium ions from the 'oxide' on platinum (111) gave almost the same pattern as a clean platinum (111) surface, although Auger electron spectroscopy confirmed the presence of oxygen. More recently it has been suggested that the stabilization of oxygen may be associated with small, and hitherto undetected, amounts of surface impurities, especially silicon (Bonzel, Franken, and Pirug 1981).

The photoelectron spectroscopy evidence for bonding is less helpful than in the case of CO. In UPS, a maximum at about 6 eV below the Fermi level is observed from the atomic chemisorbed oxygen. The emission is attributed to atomic O(2p) levels coupled to the surface. No useful distinction between the XPS of adsorbed molecular or atomic oxygen on platinum (111) could be drawn.

High resolution electron energy loss spectroscopy of oxygen on platinum (111) showed a strong band at 870 cm^{-1} when molecular oxygen was present on the surface or at 490 cm^{-1} when the oxygen was atomic. The former higher frequency band was attributed to a surface peroxo species O_2^{2-} in which metal electrons were donated into the antibonding π orbitals of O_2. Interestingly, the oxygen axis was thought to be oriented parallel to the surface, an orientation which would normally be inactive for specularly scattered electrons. The vibrational excitation is made possible by the mixing of the forbidden O—O stretch with the allowed M—O stretch. The atomic oxygen band at 490 cm^{-1} is the Pt—O stretch. It showed no frequency shift with increased coverage. The high temperature 'oxide' showed an intense absorption at 760 cm^{-1} due to coupling with the phonon modes of platinum oxide.

The analysis of LEED data is made complicated by the difficulty of obtaining reliable values of oxygen coverage. The LEED patterns themselves do not provide sufficient internal evidence as to the coverage to which they correspond, whilst reliable absolute measurements of uptakes of oxygen are notoriously difficult to make. The reasons for this difficulty are partly because the reaction of oxygen with the carbon frequently to be found on glowing filaments of ionization gauges and mass spectrometers produces CO, and partly because oxygen readily replaces CO from the walls of an ultrahigh vacuum apparatus. Oxygen also tends to make the pressure readings of conventional ionization gauges unreliable below about 10^{-7}–10^{-8} Torr.

The LEED patterns obtained from oxygen layers on the (111) planes are (2×2). The oxygen coverage is difficult to assign, but should correspond to $\theta_0 = 0.25$. On palladium (111) cooling to 200 K causes additional oxygen adsorption and gives the $(\sqrt{3} \times \sqrt{3})R30°$ pattern, which is confidently assigned to $\theta_0 = \frac{1}{3}$. For this metal, therefore, the (2×2) pattern corresponds to a coverage less than $\frac{1}{3}$ and is thus indeed due to $\theta = 0.25$. As far as the (100) and (110) planes are concerned, the spontaneous reconstruction of the platinum surfaces again causes difficulty. There is apparently a major difference in the behaviour of platinum (100) (1×1) and the reconstructed platinum (100) (5×20). The latter is extremely reluctant to adsorb oxygen under nomal circumstances whereas the former readily adsorbs oxygen to give (5×1) and (2×1) patterns. The reason for the failure to observe oxygen adsorption on platinum (100)

(5×20) has now been elucidated (Barteau, Ko, and Madix 1981). It is not that the surface is completely unreactive towards oxygen, but rather that any oxygen adsorbed on the surface is highly reactive towards CO and hydrogen. Since these gases form the major part of the residual atmosphere in an ultrahigh vacuum apparatus, and may be displaced from the walls when oxygen is admitted, any oxygen adsorbed is promptly removed even under ultrahigh vacuum conditions. By taking extreme care to avoid contamination of the admitted oxygen with CO (from the walls of the storage chamber) and reducing CO adsorptions by working at the highest temperature at which oxygen was stable on the surface, a considerable oxygen uptake on platinum (100) (5×20) was achieved. For example, using a temperature of 500 K, at which the CO adsorption is greatly reduced, the surface coverage reached about two-thirds of a monolayer of oxygen. On palladium (100) adsorption to give a (2×2) pattern was observed, whilst on palladium (110) a variety of patterns was seen. The reconstructed platinum (110) (2×1) plane adsorbed oxygen to give a (1×2) structure. In general, therefore, it can be concluded that oxygen adsorbs to give ordered surface structures on platinum surfaces.

From the point of view of the catalytic oxidation of CO there is one particularly important difference between oxygen and CO in LEED. This is the difference in the annealing temperature that is required for the patterns to develop; a much higher temperature is required for oxygen layers than for CO. Since the function of annealing is to allow the initially adsorbed gas to migrate and produce the preferred surface arrangement, we conclude that oxygen is considerably less mobile than CO. This conclusion is by no means surprising, since it is commonly, though not universally, found that the activation energy for migration is very roughly one-fifth of the surface bond strength. Since we are comparing, in round terms, bond strengths of $\sim 140 \text{ kJ mol}^{-1}$ for adsorbed CO and $\sim 360 \text{ kJ mol}^{-1}$ for adsorbed atomic oxygen, the observations accord with precedent.

2. Adsorption and desorption

Platinum and palladium react readily with oxygen, the initial sticking properties generally lying between 0.1 and 1 at room temperature. The reconstructed platinum (100) (5×20) surface discussed above may be an exception to this generalization. In view of the special reactivity of platinum (100) (5×20)—O towards hydrogen and CO it is as well to reserve judgement on the correct value of s_0 at room temperature; the lower limit of s_0 at 585 K is 1×10^{-3}, and the true value may be higher at 300 K.

The effect of introducing atomic steps onto the surface is to enhance the rate of adsorption on the platinum (111) plane. Thus on Pt(S) [12(111)×(111)], i.e. a surface consisting of terraces with the (111) arrangement which are twelve atoms wide and are connected by steps one atom high also with the (111) arrangement (see Fig. 4.7), the initial sticking probability was about 0.4, up from about 0.05 on platinum (111). However, higher values than 0.05 for s_0 of oxygen on other specimens of single-crystal platinum (111) have been reported. Sticking probability curves tend to show the influence of precursor state formation as illustrated for a polycrystalline platinum filament in Fig. 9.9. Platinum (111) is an exception to this; s_0 declines steeply with increased coverage.

The desorption of the chemisorbed atomic layer from both palladium and platinum takes place entirely by the recombination of oxygen atoms and the evolution of molecular oxygen. If a mixture of the isotopes $^{16}O_2$ and $^{18}O_2$ is adsorbed into the atomic state, complete mixing occurs. This mechanism contrasts with the observations on some high melting transition metals from which metallic oxides or oxygen atoms are desorbed following oxygen chemisorption. Although theory suggests that oxygen desorption from platinum and palladium should occur with second-order kinetics and that activation energies for desorption should be calculable, in practice this is not always so. The kinetic behaviour often appears to vary with coverage, and there is the extra complexity introduced by the surface oxide. As an example we shall consider the behaviour of oxygen

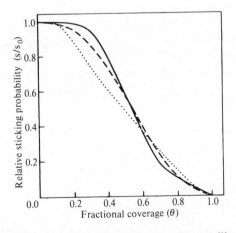

Fig. 9.9. Sticking probability curve for oxygen on polycrystalline platinum (———) compared with a precursor state theory (– – –) and an earlier similar experiment (. . . .). (From Gasser and Perry 1979).

on reconstructed platinum (100) (5×20) (Barteau *et al.* 1981). Oxygen desorption from the reconstructed surface occurred in two peaks at 676 K and 709 K, following adsorption at 585 K. The lower temperature peak could be analysed in terms of second-order kinetics, but only by including a coverage-dependent activation energy for desorption. The source of the dependence was thought to be attractive interactions between adsorbed species, since E_{des} increased with coverage. The equation for the activation energy of desorption was

$$E_{des} = 189 + \frac{30c}{c_0} \, \text{kJ mol}^{-1}$$

where c and c_0 are the instantaneous and initial coverages respectively. On the other hand, the higher temperature peak obeyed first-order kinetics with an energy of desorption of 161 kJ mol^{-1}. These values of the desorption energy and thus of the heat of adsorption are typical. Measurement of the isosteric heat for palladium (100) at medium coverage gave a value of about 242 kJ mol^{-1}, whilst on palladium (110) a variation with coverage from 335 to 200 kJ mol^{-1} was observed.

The bond strength χ of the M—O bond is related to the enthalpy of adsorption by

$$\chi = \tfrac{1}{2}(-\Delta H_{ad} + D_{O_2})$$

where D_{O_2} is the dissociation energy of oxygen which is 490 kJ mol^{-1}. Thus for the typical value of $\Delta H_{ad} = -240$ kJ mol^{-1}, the bond strength is 365 kJ mol^{-1}.

In summary, the most important results pertaining to catalytic oxidation can be gathered together as follows:

1. oxygen is dissociatively adsorbed at room temperature and above;
2. at the highest temperatures ($\geqslant 800$ K) incorporation of oxygen occurs giving a 'surface oxide';
3. oxygen is much less mobile on platinum or palladium surfaces than is CO;
4. considerable uptake of surface atomic oxygen is possible at temperatures as high as 500 K; desorption of this chemisorbed oxygen takes place in the 600–1000 K range;
5. the sticking probability on a sparsely covered surface remains considerable (up to 0.5) at catalytic temperatures.

Co-adsorption

Having thus considered the characteristics of adsorbed CO and oxygen separately we turn now to their surface interactions with each another.

Such interactions are the precursors to the catalytic formation of CO_2 and we wish to determine what happens in a mixed layer before CO_2 is evolved. The very fact that CO_2 formation occurs readily at and above room temperature on palladium and platinum complicates the enquiry, since it is necessary to inhibit the reaction by carrying out relatively low temperature studies. Under these circumstances the Langmuir adsorption model of competition for sites developed in the previous chapter is expected to be a considerable oversimplification.

The most important result established from sequential adsorption experiments on palladium is an asymmetry in behaviour between initial layers of oxygen and initial layers of CO. Thus, whereas a palladium (111) surface initially saturated with oxygen (LEED pattern (2×2), $\theta = \frac{1}{4}$) will adsorb substantial amounts of CO, the reverse is not true; when θ_{CO} exceeds $\frac{1}{3}$ (at which coverage the LEED pattern is $(\sqrt{3} \times \sqrt{3})R30°$) no dissociative adsorption of oxygen is possible. The effect of CO adsorption

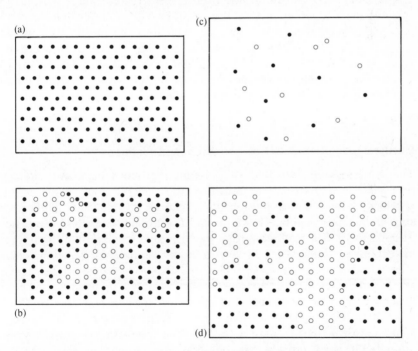

Fig. 9.10. The compression of an $O(2 \times 2)$ layer by subsequent adsorption of CO: (a) $O(2 \times 2)$ on palladium (111); (b) domains of $O(\sqrt{3} \times \sqrt{3})R30°$ (●) and CO $(\sqrt{3} \times \sqrt{3})R30°$ (○) on palladium (111). Growth of ordered structures from the random arrangement at low coverage to the preferred structure for each adsorbate at higher coverages: (c) O and CO randomly adsorbed at low coverage on palladium (111); (d) domains of $O(2 \times 2)$ (●) and $CO(\sqrt{3} \times \sqrt{3})R30°$ (○) on palladium (111). (From Conrad, Ertl, and Küppers 1978.)

on an existing oxygen layer is to cause compression of the original $O(2\times2)$ layer structure ($\theta_0 = \frac{1}{4}$) shown in Fig. 9.10(a) to $O(\sqrt{3}\times\sqrt{3})R30°$ ($\theta_0 = \frac{1}{3}$), the compression weakening the bond between oxygen and the metal. The CO is then able to adsorb on the space thus made available, giving the $CO(\sqrt{3}\times\sqrt{3})R30°$ structure. The palladium (111) surface is thus covered by two sets of interlocking but separate lattices, rather like a jigsaw puzzle, as illustrated in Fig. 9.10(b). A similar territorial separation is observed if the initial CO coverage is less than $\frac{1}{3}$, i.e. not so large as to inhibit totally the subsequent adsorption of oxygen. At sufficiently low initial oxygen and CO coverages, both CO molecules and O atoms are distributed at random. As the coverages increase the O atoms take up their preferred $O(2\times2)$ arrangement whilst the CO molecules adopt the $CO(\sqrt{3}\times\sqrt{3})R30°$ structure. The development of these structures is illustrated in Figs. 9.10(c) and 9.10(d).

Two inferences can be drawn from these structural data. The first is that on the palladium surface the O atom–CO molecule interaction is energetically unfavourable compared with the O–O or CO–CO interactions. The second is that the domains on the surface over which the separate ordered arrays extend are greater than ~ 100 Å, the coherence width (see Chapter 5) of the electrons in this experiment. Platinum surfaces are expected to give rise to similar behaviour.

Catalytic oxidation of CO on palladium and platinum

The experimental fact that the platinum group metals are effective catalysts for CO oxidation illustrates the general point made in the previous chapter that a good catalyst adsorbs the reactants neither too strongly nor too weakly. Thus with an M—O bond strength in the region of 360 kJ mol^{-1}, a platinum group metal is able to dissociate the oxygen molecule and produce the surface O atoms required for reaction. These atoms, however, are not so strongly bonded as to be rendered unreactive. We can thus contrast on the one hand the catalytic unreactivity of gold, which can just about be induced to adsorb oxygen (~ 1 Torr of O_2 at $\sim 200\,°C$) with on the other hand the catalytic unreactivity of tungsten which adsorbs oxygen 'strongly' (a bond strength of about 600 kJ atom^{-1}). It is of interest to note that on thermodynamic criteria, a potential candidate for a CO oxidation catalyst is silver, and indeed a layer of oxygen on silver does readily react with CO. However, the sticking probability of oxygen on silver is so low that once the oxygen layer has been removed it is not replenished. Steady state oxidation is therefore negligible. These observations illustrate the obvious but important considerations for effective catalysis, that not only must the thermodynamic factors be favourable but also the primary adsorption process must not be slow.

General considerations

The main aim of low pressure measurements of the oxidation of CO in the presence of platinum or palladium has been to elucidate the mechanism. The discussion has centred on the basic choice between the Eley–Rideal and the Langmuir–Hinshelwood mechanisms. The evolution of opinion as to the most satisfactory interpretation of the results illustrates well how difficult it is in practice to reach a confident conclusion. Indeed, although, as we shall see, the weight of the current evidence favours the Langmuir–Hinshelwood mechanism on palladium, it is only a few years since the Eley–Rideal mechanism seemed to give the more satisfactory overall explanation. As far as platinum is concerned it is clear that under most circumstances the Langmuir–Hinshelwood is the predominant mechanism. Indeed, some authors have proposed that this is the only mechanism. However, the evidence on this point is conflicting, although the most recent work appears to favour the exclusive view.

It is of some interest to enquire why it is so difficult to assign a mechanism to even such a simple surface process as catalytic oxidation of CO, when much more complicated reactions in solution seem to be well understood. A substantial part of the reason must rest with the use of the Langmuir isotherm (although there is no reason to be optimistic that the alternatives would be superior) for estimating surface coverages.

Let us consider the assumptions implicit in this equation and its use for a system in which a heterogeneous reaction is occurring.

In the first place, the isotherm is derived with the assumption of dynamic equilibrium between the gas and the surface. For the surface reaction to be rate determining, this equilibrium must be established rapidly compared with the rate of reaction. If it is not, the rate of adsorption or desorption of one of the reactants may become the rate-determining step. In addition, when the product is strongly adsorbed, its rate of desorption may become limiting.

A second fundametal difficulty is apparent from the earlier discussions of adsorption and co-adsorption. In adsorption, the Langmuir assumption that the surface is energetically homogeneous may be unjustified. Or, even if it is true at low coverage, as might be hoped in the example of a single-crystal plane, adsorption may induce heterogeneity. We saw an example of this process for CO on palladium (111), where changes in the heat of adsorption were brought about as θ_{CO} increased. Molecules which had been adsorbed at lower coverages were constrained by subsequent adsorption to move to sites with lower binding energy.

Turning now to co-adsorption, the idea of a competition for places on the surface based solely on the heats of adsorption and the gas pressures of the individual adsorbates is too simple. Thus the adsorption of one species may drastically alter the binding energy of the other, and in a way which depends on the coverage of both. The adsorption of CO on oxygen-covered palladium (111) is an example of just such a process where the O atoms become compressed and their bonding is weakened. This process is bound to have a profound affect on the activation energy for the surface reaction between CO and oxygen and thus on its rate. Therefore it is with these caveats in mind that we turn now to a consideration of the evidence.

Evidence for the mechanism of CO catalytic oxidation

An interpretation of the experimental evidence in terms of either the Langmuir–Hinshelwood or the Eley–Rideal mechanism requires a choice between the following possibilities.

(a) *Langmuir–Hinshelwood mechanism*:

$$CO(g) \xrightarrow{k_1} CO(ad) \tag{1}$$

$$CO(ad) \xrightarrow{k_2} CO(g) \tag{2}$$

$$O_2(g) \xrightarrow{k_3} 2O(ad) \tag{3}$$

$$CO(ad) + O(ad) \xrightarrow{k_4} CO_2(g). \tag{4}$$

(b) *Eley–Rideal Mechanism*

Either

$$O_2(g) \xrightarrow{k_3} 2O(ad)$$

$$O(ad) + CO(g) \xrightarrow{k_5} CO_2(g) \tag{1}$$

or

$$CO(g) \underset{k_2}{\overset{k_1}{\rightleftharpoons}} CO(ad)$$

$$2CO(ad) + O_2(g) \xrightarrow{k_6} 2CO_2(g). \tag{2}$$

In the Eley–Rideal mechanism the non-adsorbed reactant is assumed to react either directly during collision from the gas phase or via a transient very weakly physisorbed molecular surface species. The surface coverage of this latter species is miniscule, so that limiting low pressure Langmuir behaviour, ($\theta \propto P$) can be assumed to be followed (see Chapter 1). The overall rate is then proportional to $\theta_0 p_{CO}$ (eqn 8.5).

One fortunate simplying feature of the reaction is that CO_2 is not adsorbed by either platinum or palladium. There is, therefore, no question of inhibition by the reaction product. Furthermore, the second Eley–Rideal possibility (b) (2) can immediately be ruled out. A saturated layer of CO on platinum or palladium does not react with oxygen. We are thus left with a choice between (a) (1)–(4) and (b) (1).

Many of the experimental methods used to try and decide between these alternatives fall into one of two categories. In the first group are steady state experiments in which the metal is exposed to a mixture of oxygen and CO and the rate of production of CO_2 is monitored mass spectrometrically. The most important variables are the temperature of the metal, the individual gas pressures p_{CO} and p_{O_2}, and the ratio of the partial pressures p_{O_2}/p_{CO}. The second kind of experiment establishes the system at a condition far from the steady state and follows the subsequent rate of reaction, again mass spectrometrically. A typical non-steady state procedure is to adsorb one or both of the gases under chosen conditions of temperature and pressure. The pressures of the gases are then altered abruptly and the subsequent evolution of CO_2 is recorded. Broadly speaking, the latter, i.e. non-steady state, experiments lead to mechanisic conclusions by a consideration of the temperature dependence of the reaction rate. On the other hand, pressure dependences as well as temperature dependences of rates can usefully be discussed under steady state conditions. Some results for both types of experiment will be considered. Before doing so the modulated molecular beam technique, which is a particularly valuable example of the use of non-stationary state methods, will be discussed. It has been claimed to give the most direct evidence for the reaction mechanism (9.1). We shall then consider how far other results are consistent with this proposed mechanism.

Modulated molecular beam results

In this technique one of the reactive gases is in the form of a molecular beam whose intensity can be modulated. The other gas is at a constant pressure and undergoes isotropic collisions with the surface.

For the reaction catalysed by palladium (111), a beam of either CO or oxygen was allowed to impinge on the single-crystal surface, typically at rate equivalent to a pressure of 10^{-7} Torr, the other gas being at an isotropic pressure of the same order. A shutter was then oscillated in and out of the path of the beam at a frequency in the $100-1000\ s^{-1}$ range, thus producing a modulated beam. The CO_2 produced was measured using a rapid response mass spectrometer located in the ultrahigh vacuum chamber so that it was in line of sight from the palladium surface. CO_2 molecules were thus detected before they had undergone any collisions with the walls. Because of the oscillation in the pressure of molecular beam gas at the surface there was a corresponding oscillation in the rate of production of CO_2 and thus in the amplitude of the mass spectrometer signal.

There is, of course, a time lag between the opening of the shutter in the molecular beam and the appearance of CO_2 at the detector. The delay is determined in part by the travelling times of CO molecules from the shutter to the crystal and of CO_2 molecules from the crystal to the detector. These times are not important to the discussion. What is important from a kinetic point of view is the additional delay occasioned by the time taken for the reaction to occur on the surface. The slower is the reaction the greater is the delay. Measurement of this delay time thus gives access to the rate of surface reaction. In practice, a convenient method of measurement is by way of phase sensitive detection. The relevant feature of this technique is that the experimentally inconvenient time delay is converted to the readily measured phase angle difference between the oscillations of the shutter and the oscillations of the CO_2 mass spectrometer signal; the longer is the reaction time the greater is the phase-lag angle ϕ.

The results from a CO-modulated beam with O_2 as the background gas will be discussed first.

The experimental conditions chosen were as follows.

1. $700\ K > T > 500\ K$: at these temperatures the oxygen is on the surface in the atomic state and the coverage of CO is low ($\theta_{CO} < 0.03$).

2. $\theta_O > 0.1$, which ensures a considerable excess of oxygen over CO on the surface. This condition has two advantages: (a) it ensures that depletion of oxygen during exposure of the surface to CO during the open part of the shutter cycle is negligible, which simplifies the analysis of results, and (b) the rate of reaction is not determined by oxygen adsorption.

The rate of production of CO_2 depends upon the surface coverage of oxygen $f(\theta_O)$ (ideally $f(\theta_O) = \theta_O$, but as we saw above this is too simple a function) and either the surface coverage (Langmuir–Hinshelwood) or the pressure (Eley–Rideal) of CO, i.e.

$$\frac{d[CO_2]}{dt} = k_4 \theta_{CO} f(\theta_O) \qquad \text{Langmuir–Hinshelwood}$$

or

$$\frac{d[CO_2]}{dt} = k_5 p_{CO} f(\theta_O) \qquad \text{Eley–Rideal.}$$

If now the pressure of CO is modulated at a frequency ω according to the equation

$$p_{CO} = p_{CO}^0 \alpha \exp(i\omega t),$$

where α is the amplitude of modulation, the Langmuir–Hinshelwood mechanism predicts that the rate of reaction oscillates according to the equation

$$\frac{d[CO_2]}{dt} = k_4 f(\theta_O)\{\theta_{CO}^0 + f(LH)\} \qquad (9.1)$$

where the Langmuir–Hinshelwood function $f(LH)$ is complex and includes ω, k_1, k_2 and $k_4 f(\theta_O)$. For this mechanism the phase-lag angle ϕ is given by

$$\tan \phi = \frac{\omega}{k_2 + k_4 f(\theta_O)}. \qquad (9.2)$$

The Eley–Rideal mechanism gives the rate of reaction as

$$\frac{d[CO_2]}{dt} = k_5 f(\theta_O)\{p_{CO}^0 + \alpha \exp(i\omega t)\} \qquad (9.3)$$

and

$$\tan \phi = 0.$$

The important features of eqns (9.2) and (9.3) are that, whereas in the Langmuir–Hinshelwood case $\tan \phi$ is expected to be dependent on temperature (through either or both of k_2 and k_4), for the Eley–Rideal mechanism a temperature-independent ϕ is predicted. In physical terms we can see why these results are obtained. The phase lag is fundamentally due to the finite time taken for the reaction between a colliding CO molecule and the surface O atom. In the Langmuir–Hinshelwood case, where the CO molecule becomes adsorbed on the surface, an increase in temperature increases the rate of this reaction. The characteristic time to produce a CO_2 molecule (also known as the relaxation time and given by $\tau = 1/k$) is thereby reduced and thus ϕ decreases. On the other hand in the Eley–Rideal mechanism the CO_2 molecule is produced during the collision and without delay. Increase in temperature therefore does not affect ϕ since this is in any case zero.

Returning now to the Langmuir–Hinshelwood equation (eqn (9.2)) we note that in the denominator k_2 describes the rate of desorption of surface CO and $k_4 f(\theta_O)$ describes its rate of reaction. In the situation where surface reaction is the rate-determining step $k_2 \gg k_4 f(\theta_O)$. This situation obtains at $T > 500$ K. Thus we can simplify eqn (9.1) to give

$$\tan \phi = \omega/k_2 \qquad (9.4)$$

An Arrhenius-type plot of $\ln(\omega/\tan \phi)$, i.e. $\ln k_2$, versus $1/T$ is thus predicted to be a straight line of slope $(-E^*/R)$, where E^* is the activation energy of reaction (a) (2). For non-activated adsorption, E^* is equal to the heat of adsorption of CO on a surface with the particular coverage of oxygen obtaining during the experiment. Equation (9.4) also shows that for a very weak adsorption with a low value of E^*, $\tan \phi \rightarrow 0$ since then $k_2 \gg \omega$. This is what happens in the Eley–Rideal mechanism if the reaction is postulated to occur between a very weakly adsorbed CO molecule and a surface O atom. Again, therefore, a temperature-independent value of ϕ is predicted for an Eley–Rideal reaction.

The results are plotted in Fig. 9.11, where the relaxation time ($\tau = 1/k_2$), rather than the velocity constant, has been chosen as the abscissa.

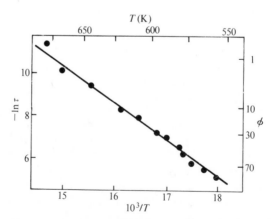

Fig. 9.11. Arrhenius plot of the relaxation time ($\tau = 1/k_2$) for CO oxidation on palladium (111) vs. $1/T$. The CO beam was modulated at $\omega = 779 \, \text{s}^{-1}$. The phase lag angle ϕ from which τ is calculated is also shown. (From Engel and Ertl 1978.)

An excellent straight line is obtained whose slope gives a value of $E^* = 139\,\text{kJ mol}^{-1}$. This value is in agreement with the isosteric heat of adsorption ($143\,\text{kJ mol}^{-1}$) and with values obtained from rate measurements. It is clear therefore that these results point unambiguously to a Langmuir–Hinshelwood mechanism under the conditions specified. However, the temperature dependence of ϕ does not give information about the activation energy E_{LH} for the surface reaction. To obtain this quantity we return to eqn (9.1). The Langmuir–Hinshelwood function $f(\text{LH})$ can be simplified, under the conditions of this experiment ($k_2 \gg k_4 f(\theta_0)$), to the form

$$J_{\text{CO}_2} = \frac{\text{const. } k_4}{(k_2^2 + \omega^2)^{1/2}}$$

where J_{CO_2} is the amplitude of the mass spectrometer signal for CO_2. In this equation k_2 and ω are known, so if we assume that k_4 has the usual Arrhenius form $k_4 = \nu_4 \exp(-E_{\text{LH}}/RT)$, the shape of the curve for the temperature dependence of J_{CO_2} can be calculated. As Fig. 9.12 shows, a good fit is obtained for $E_{\text{LH}} = 105\,\text{kJ mol}^{-1}$. In summary, it can be concluded that for the reaction at low coverage of CO and a considerable excess of adsorbed oxygen, the modulated CO beam results point clearly to the Langmuir–Hinshelwood mechanism.

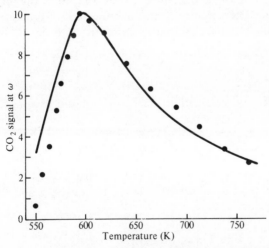

Fig. 9.12. Catalytic oxidation by palladium (111). Comparison of the mass spectrometer signal J_{CO_2} for CO_2 as measured experimentally (●) with values calculated assuming $E_{\text{LH}} = 105\,\text{kJ mol}^{-1}$ (———). The CO beam was modulated at $\omega = 779\,\text{s}^{-1}$. (From Engel and Ertl 1978.)

Turning now to the case of a modulated beam of oxygen molecules and CO as the background gas the experimental conditions were as follows.

1. $475\,\text{K} < T < 550\,\text{K}$; the surface oxygen is again in the atomic state.

2. $\theta_{CO} \gg \theta_O$ at all times during the reaction so that θ_{CO} did not change appreciably during the reaction period.

3. the rate of reaction, as established by separate experiments, was directly proportional to θ_O.

The rate equation can now be formulated as

$$\frac{d[CO_2]}{dt} = k_4 \theta_O \theta_{CO} \qquad \text{Langmuir–Hinshelwood}$$

or

$$\frac{d[CO_2]}{dt} = k_5 \theta_O p_{CO} \qquad \text{Eley–Rideal.}$$

When the oxygen pressure is modulated, the phase-lag angles are given by

$$\tan \phi_{LH} = \frac{\omega}{k_4 \theta_{CO}} \qquad \text{Langmuir–Hinshelwood} \qquad (9.5)$$

$$\tan \phi_{ER} = \frac{\omega}{k_5 p_{CO}} \qquad \text{Eley–Rideal.} \qquad (9.6)$$

The temperature dependences of the phase-lag angles are of opposite sign and thus provide an immediate diagnostic for the mechanism. We can see why this is so by noting that at temperatures near 500 K the coverage of CO is low and there is indeed dynamic equilibrium between gas and surface. The conditions for Langmuir adsorption are thus reasonably well fulfilled and we can write in accord with eqn (1.6) for $\theta \ll 1$

$$\theta_{CO} = p_{CO}(2\pi m k T)^{-1/2} \frac{s^*}{k_d} \exp\left(\frac{-\Delta H_{ad}}{RT}\right). \qquad (9.7)$$

For non-activated adsorption $-\Delta H_{ad} = E_{des}$, and at constant pressure

$$\theta_{CO} \propto \exp\left(\frac{E_{des}}{RT}\right). \qquad (9.8)$$

If now the rate constants k_4 and k_5 are expressed in the usual Arrhenius form and substitution is made in eqn (9.5) and (9.6) together with eqn (9.8), the result is

$$\tan \phi_{LH} \propto \exp\left(\frac{E_{LH} - E_{des}}{RT}\right)$$

$$\tan \phi_{ER} \propto \exp\left(\frac{E_{ER}}{RT}\right). \qquad (9.9)$$

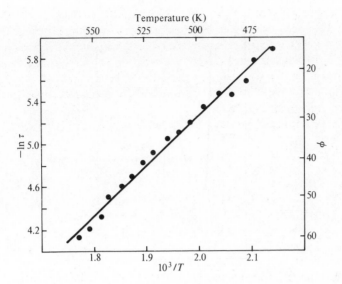

Fig. 9.13. Catalytic oxidation by palladium (111). Arrhenius plot of the relaxation time τ for CO oxidation on palladium (111) vs. $1/T$. The oxygen beam was modulated at $113\,s^{-1}$. The phase lag angle ϕ is also shown. (From Engel and Ertl 1978.)

The previous beam experiment showed that $E_{des} > E_{LH}$ ($139\,kJ\,mol^{-1}$ and $105\,kJ\,mol^{-1}$). Thus ϕ_{LH} starts at a low value (in the limit zero) at low temperature and increases with temperature (in the limit to $\pi/2$), whereas the opposite is true for ϕ_{ER}. The variation of ϕ with T shown in Fig. 9.13 is clear evidence for the Langmuir–Hinshelwood mechanism. Also shown in Fig. 9:13 is the temperature dependence of the relaxation time τ calculated from the phase-lag angle. The slope of the line in Fig. 9.13 yields $E_{LM} - E_{des}$ as $-40\,kJ\,mol^{-1}$, in accordance with eqn (9.9). If now E_{des} is equated with the isosteric heat of adsorption of CO ($143\,kJ\,mol^{-1}$) then E_{LH} is $-40 + 143 = 103\,kJ\,mol^{-1}$ which is in excellent agreement with the previous value of $105\,kJ\,mol^{-1}$.

It seems safe to conclude that under the conditions of the two beam experiments, and it should be noted that the coverages and temperature ranges have been carefully selected, the catalytic oxidation of CO on palladium (111) proceeds via the Langmuir–Hinshelwood mechanism.

The results of a subsequent molecular beam study of the catalytic oxidation of CO on a platinum (111) surface were similar to those just described for palladium (111). (Campbell, Ertl, Kuipers, and Segner

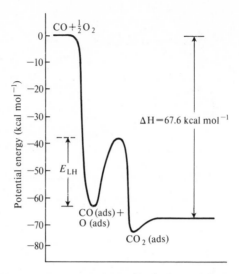

Fig. 9.14. Catalytic oxidation by platinum (111). Potential energy diagram for the catalytic mechanism for the low coverage limit. (From Campbell *et al.* 1980.)

1980). Under all the conditions studied the reaction proceeded by the Langmuir–Hinshelwood mechanism, with an activation energy for reaction which decreased from 101 kJ mol^{-1} at low coverages to 49 kJ mol^{-1} at high oxygen coverage. The potential energy profile for the reaction taking place at low coverage is illustrated in Fig. 9.14. The reduction in activation energy at higher coverages was attributed, as before, to repulsion between CO_{ad} and O_{ad}. The result of this effect would be to raise the $CO_{ad} + O_{ad}$ level in Fig. 9.14, since the overall enthalpy of adsorption from the $CO + \frac{1}{2}O_2$ level is reduced, and thus reduce E_{LH}. This interpretation is similar to that given for the 'strong' adsorption side of the volcano curve discussed in Chapter 8. It is of interest to note that most of the enthalpy of the reaction $CO + \frac{1}{2}O_2 \rightarrow CO_2$ is liberated in the primary adsorption process ($E_{ad} = 146 \text{ kJ mol}^{-1}$ for CO at $\theta_{CO} = 0$ and $E_{ad} = 115 \text{ kJ mol}^{-1}$ for $\frac{1}{2}O_2$ at $\theta_O = 0$). The surface reaction is only mildly exothermic.

The kinetic data were broadly similar to the observations on palladium (111) and were of equal complexity.

Other non-steady state results

1. Palladium

The first set of transient phenomena we shall discuss continue the molecular beam story of the catalytic activity of palladium (111). The

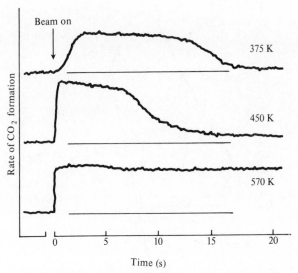

Fig. 9.15. Catalytic activity of palladium (111). Non-steady state rate of CO_2 production as a function of time for a CO beam (equivalent pressure 6×10^{-8} Torr) impinging on the substrate for three substrate temperatures ($p_{O_2} = 1 \times 10^{-7}$ Torr). (From Engel and Ertl 1978.)

surface was initially saturated with oxygen (LEED pattern (2×2, $\theta_O = \frac{1}{4}$) and maintained in a pressure of 1×10^{-7} Torr of the gas. The CO beam was then admitted to give an equivalent pressure of 6×10^{-8} Torr. The evolution of CO_2 was followed until steady state conditions were reached. Some typical results are shown in Fig. 9.15. As can be seen there was some delay before the rate of production of CO_2 reached its maximum value; the lower was the catalyst temperature the longer was the delay. The later steady state stages of the reaction also depended on the temperature. At lower temperatures ($T < 500$ K) the steady state rate of evolution was less than the maximum rate, whereas at higher temperatures ($T > 550$ K) the two rates were identical. A comparison of the maximum and steady state rates of CO_2 production for 350 K $< T <$ 650 K is shown in Fig. 9.16.

A number of qualitative deductions can be made from these results. The first is that the reaction is not proceeding by an Eley–Rideal mechanism between CO(g) and O(ad). If it were, the maximum rate would be expected immediately on the admission of CO, contrary to observation. Consonant with this reasoning we may note that below T_{max} (550 K) some delay in the maximum rate of production of CO_2 is expected in the Langmuir–Hinshelwood reaction as CO builds up on the surface, as observed. Secondly, as we have already seen, adsorption of

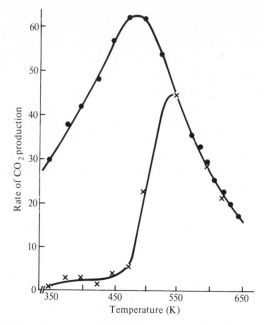

Fig. 9.16. Catalytic activity of palladium (111). A comparison of the maximum and the steady state rates of CO_2 production obtained from curves of the type shown in Fig. 9.15 at various substrate temperatures: ●, r_{max}; ×, $r_{steady\ state}$. (From Engel and Ertl 1978.)

CO on an $O(2 \times 2)$ layer causes compression and weakening of the Pd—O bond. It is not unexpected therefore that in the lower temperature region ($T < 550$ K) the maximum rate of production of CO_2 is more rapid on an oxygen-rich surface than under steady state conditions; the activation energy is, after all, lower. Finally, we note that at higher temperatures ($T > 550$ K) no difference is anticipated, or found, between the maximum and steady state rates. At $T > 550$ K the equilibrium coverage of CO is so low that there is not enough surface CO to affect the binding energy of the oxygen atoms.

Having again postulated reaction via the Langmuir–Hinshelwood mechanism the next step is to seek a method of analysing the rate data to obtain a value for the activation energy E_{LH}. Before so doing, we should note that in writing the rate equation for a Langmuir–Hinshelwood reaction as

$$R = k_{LH}\theta_O\theta_{CO}$$

the velocity constant k_{LH} is a complex function of θ_O and θ_{CO}. Indeed, the above discussion led us to suppose that E_{LH} was significantly different above and below T_{max}. Below T_{max}, the compression of the atomic oxygen layer by adsorbed CO weakens its bonding to the surface and thus reduces E_{LH}. Above T_{max} the equilibrium uptake of CO is miniscule and a higher value of E_{LH} is expected.

In order to put these considerations on a quantitative footing, we turn to experiments in which the rates of CO_2 production on palladium (111) at different temperatures and at known values of θ_O and θ_{CO} were compared. For this work it was experimentally convenient to switch off the oxygen supply before admitting the CO. Analysis of the CO_2 evolution curves then allowed the surface composition to be determined as the oxygen was removed. One example of the way in which θ_O, θ_{CO}, and R_{CO_2} varied with time at $T < T_{max}$ when θ_{CO} was considerable is shown in Fig. 9.17(a). At higher temperatures ($T > T_{max}$) θ_{CO} is always low and the curves look like those illustrated in Fig. 9.17(b). Curves such as those in Fig. 9.17 were used to derive the Arrhenius plot illustrated in Fig. 9.18 for estimating E_{LH}. The values of the activation energies deduced from the two straight lines in Fig. 9.18 were

$$E_{LH} \text{ (low temperature, considerable } \theta_{CO}) = 59 \text{ kJ mol}^{-1}$$

$$E_{LH} \text{ (high temperature, low } \theta_{CO}) = 113 \text{ kJ mol}^{-1}.$$

The high temperature value of E_{LH} is in satisfactory agreement with the modulated molecular beam result (i.e. within the experimental error), whilst the low temperature value shows the effect of compression of the oxygen layer.

2. Platinum

We now turn to a consideration of the catalytic activity of platinum and start with the results from some non-modulated beam studies of the catalytic activity of a polycrystalline platinum ribbon.

Three groups of experiments were performed: mixed-beam experiments in which both oxygen and CO were in the beam, and two single-beam experiments in which one gas formed the beam and the other was at an isotropic pressure. Gas pressures were in the range 10^{-6}–10^{-8} Torr, or equivalent in the case of the beams, and the catalyst temperature range was 300–1250 K. The first experiments with a mixed beam of CO and oxygen were used to identify the temperature dependence of the steady state production of CO_2. The reaction became measurable just below 500 K, passed through a maximum at about 700 K, and declined to zero at about 1250 K.

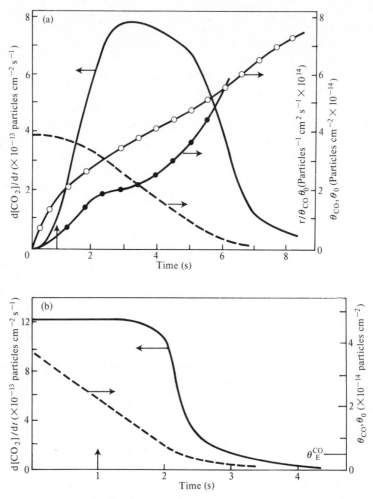

Fig. 9.17. Catalytic oxidation by palladium (111) at (a) 374 K and (b) 509 K showing the rate of CO_2 production r (———), θ_O (- - -), and θ_{CO} (O) as a function of time; ($\bullet$) shows $r/\theta_O\theta_{CO}$ as a function of time. θ_E^{CO} is the equilibrium coverage at the CO pressure and sample temperature. (From Engel and Ertl 1978.)

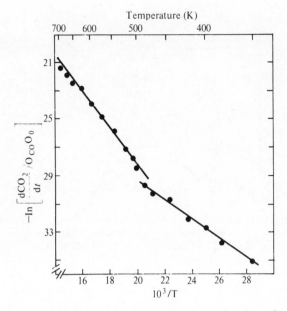

Fig. 9.18. Catalytic oxidation by palladium (111). Ln k, where $k = r/\theta_O\theta_{CO}$, as a function of $1/T$. The data were obtained from curves of the type shown in Fig. 9.17. θ_E^{CO} was used when θ_{CO} could not be measured directly. (From Engel and Ertl 1978.)

When an oxygen beam was first admitted to the polycrystalline platinum at a temperature below 700 K and in an isotropic pressure of CO, the reactive sticking probability was low but increased with time to its stationary state value. At high temperatures the stationary value was reached almost immediately. These results were interpreted using the Langmuir–Hinshelwood mechanism as follows. As long as the platinum temperature was high enough to allow any desorption of CO, the sites on the surface thus vacated could be occupied by O atoms from the beam instead of by CO from the isotropic gas. Reaction between O atoms and CO molecules was then able to occur on the surface, with the evolution of CO_2. More spaces were thus produced, more oxygen could be adsorbed, and more CO_2 could be desorbed. The rate therefore rose until eventually the steady state coverages of CO molecules and O atoms were achieved. The higher was the temperature the more rapidly was the steady state reached. The Eley–Rideal mechanism was ruled out because the maximum rate would be expected for this mechanism when the CO coverage was at its maximum, i.e. initially.

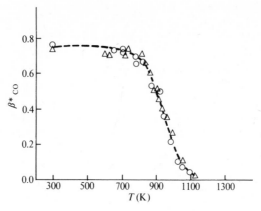

Fig. 9.19. Catalytic oxidation by polycrystalline platinum. Variation of the initial reaction probability β^*_{CO} with temperature for two different beam intensities and constant oxygen pressure. (From Pacia, Cassuto, Pentenero and Weber 1976.) $p_{O_2} = 7 \times 10^{-7}$ Torr. Equivalent $p_{f(CO)} = 2.2 \times 10^{-7}$ Torr ($\bigcirc$) or 4.2×10^{-7} Torr ($\triangle$). p_f is the isotropic pressure which would give the same collision rate as the beam.

Rather different results were obtained for the second molecular beam experiment with CO in the beam and oxygen as the isotropic gas. First, the initial rate of reaction was not low; in contrast, at room temperature the initial reactive sticking probability was 0.5–0.7. The second main point of difference lay in the temperature dependence of the initial rate, which is shown in Fig. 9.19. The initial reaction probability β^* is constant up to about 800 K, at which temperature oxygen begins to desorb. The independence of β^* from temperature between 300 and 800 K suggests that the reaction is non-activated. Furthermore, the initial rate of reaction was first order in CO pressure. Qualitiatively, at least, these results imply that the reaction is taking place through the Eley–Rideal mechanism.

An analysis of the data in the Langmuir–Hinshelwood regime ($T > 600$ K, oxygen molecular beam) gave an activation energy for this mechanism of 93 kJ mol^{-1}. As far as the Eley–Rideal experiments were concerned (300 K $> T > 1200$ K, CO molecular beam) the activation energy was zero, the temperature dependence above 900 K being attributed to a reducing oxygen coverage. Quantitative analysis of the rate data for $T > 900$ K gave an estimate of the heat of desorption of molecular oxygen as 100–140 kJ mol^{-1}. This result agrees well with independent measurements of the activation energy for desorption of oxygen and adds support for the proposed Eley–Rideal mechanism. It thus appeared, not unreasonably, that the reaction mechanism depended on the experimental conditions.

Other non-stationary state measurements on platinum have confirmed the inhibiting effect of adsorbed CO on catalytic activity at temperatures below T_{max}. Thus, for example, when a platinum (110) surface at temperatures between 390 and 450 K was exposed to a mixture of CO and oxygen the reaction rate was strongly inhibited at a sufficiently high p_{CO}/p_{O_2} ratio. The interpretation given was that at a sufficient excess of gaseous CO, the surface coverage θ_{CO} exceeded some criticla value θ_C, at which the probability of oxygen adsorption fell drastically. When the CO pressure was abruptly reduced it was possible for θ_{CO} to fall below θ_C, for oxygen to adsorb and for CO_2 to be evolved. Some time delay occurred, however, before the evolution of CO_2, whilst these processes occurred; the lower was the temperature the longer was the delay. The results are shown in Fig. 9.20. This strongly inhibiting effect of CO was thought to rule out the Eley–Rideal mechanism under these conditions.

Another time-dependent measurement of the oxidation of CO on polycrystalline platinum used an a.c. pulsing technique. For this experiment one gas was maintained at a steady background pressure whilst the pressure of the other gas was modulated, usually at ~ 5 Hz. There are some resemblances here to the modulated-beam experiments described

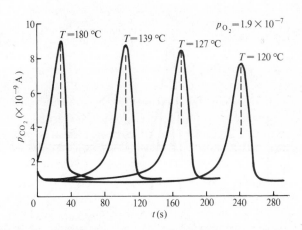

Fig. 9.20. Catalytic activity of platinum (110). Rate of CO_2 formation on platinum (110) under non-stationary conditions. The surface was initially exposed to a mixture of CO and O_2. At $t = 0$ the CO pressure was abruptly reduced and the subsequent rate of CO_2 formation was registered. The time corresponding to the maximum rate shifts to lower values with increasing surface temperature. (From Bonzel and Ku 1972.)

earlier, in that the rate of production of CO_2 had an oscillating component at the modulating frequency and this component was separated from the background CO_2 by phase-sensitive detection. Differences lie in the method of modulation of gas pressure via a piezoelectric valve to the reservoir in the present experiments, and in the fact that in this work the platinum catalyst experienced an isotropic pressure of both gases, rather than one isotropic and one beam gas. The platinum temperature was raised slowly from 400 to 900 K and the oscillating component of the CO_2 partial pressure was recorded. CO and oxygen were used in turn as the constant and modulated components. Some typical results at pressures of 5×10^{-8} Torr are shown in Fig. 9.21. The results were analysed in terms of the Langmuir–Hinshelwood mechanism. An exact solution for the temperature dependence of the a.c. component is not available algebraically although it can be computed. However, for some limited ranges of conditions approximate solutions of the rate equations were possible. These showed that when the oxygen pressure was pulsed, the modulated CO_2 signal was determined primarily by the condition that the rates of CO adsorption and desorption should be comparable. Thus, as in other work, CO desorption was the rate-determining step. On the other hand when CO was pulsed, the onset of the oscillating CO_2 signal was linked to the modulation frequence ω and occurred when this quantity

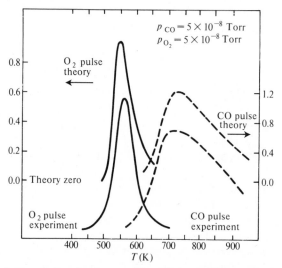

Fig. 9.21. Catalytic activity of polycrystalline platinum. Comparison of the oxygen and CO curves calculated with best-fit parameters with experimental curves for CO and oxygen pressures of 5×10^{-8} Torr. (From Strozier 1979.)

was comparable with the CO desorption rate: $k_2 \approx \omega$. Oxygen desorption was not a factor, since oxygen was stable on the surface below 900 K.

Computer fitting to the experimental curves allowed values for the rate constant parameters in the Langmuir–Hinshelwood mechanism to be deduced. A choise of $(E_{des})_{CO} = 133 \text{ kJ mol}^{-1}$ and $E_{LH} = 63 \text{ kJ mol}^{-1}$ gave the excellent fit illustrated in Fig. 9.21.

Having thus discussed some of the evidence concerning time-dependent processes on palladium and platinum we shall see how the results can be compared with steady state measurements, and we turn to these next.

Steady state oxidation

In discussing the results of steady state oxidation measurements we shall have the advantage of being able to consider both the temperature and the pressure dependence of the rate of CO_2 production. A proposed mechanism should then be consonant with both sets of observations.

1. Temperature dependence of oxidation rate

The general features are similar for palladium and platinum. There is a rise of the rate from a low level at room temperature to a maximum value which depends on the experimental conditions, but which lies between 450 and 700 K. The rate then declines, and in the case of platinum where it has been followed to 1200 K, actually becomes zero at the highest temperatures. Some results for palladium (111) are shown in Fig. 9.22. Results for polycrystalline platinum, whose non-steady state behaviour is shown in Fig. 9.19, together with the observations of platinum (110) are shown in Fig. 9.23.

No single factor accounts for the observed maxima in the steady state rate curves and the dependences on the pressures of CO and oxygen seen in Figs. 9.22 and 9.23. In interpreting the results we have to take account of the following considerations:

1. a layer of adsorbed CO inhibits CO_2 formation, whereas a layer of oxygen does not;

2. the saturation uptake of CO declines on both metals as the temperature is raised, and has become low by 600 K;

3. oxygen is thermally stable on the surfaces at temperatures where the CO coverage is low;

4. oxygen desorption occurs for $700 \text{ K} < T < 1000 \text{ K}$.

In broad terms, the rising part of the curves $(400 \text{ K} < T < 700 \text{ K})$ is attributed to CO desorption and the reduction of the inhibition mentioned in (1) above. The falling part $(T > 450\text{–}600 \text{ K})$ is attributed to the

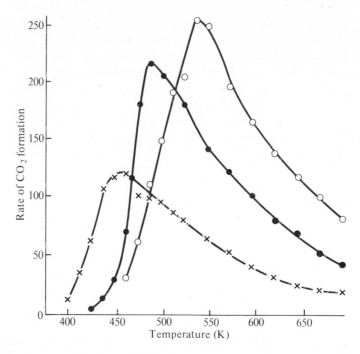

Fig. 9.22. Steady state oxidation of CO on palladium (111). Rate of CO_2 production as a function of the substrate temperature for an oxygen beam (equivalent pressure 4×10^{-7} Torr) and CO pressures of 1×10^{-7} Torr ($\times$), 3×10^{-7} Torr ($\bullet$), and 1×10^{-6} Torr ($\bigcirc$). (From Engel and Ertl 1978.)

balance between an increased Langmuir–Hinshelwood velocity constant and the reduction in what is by now a sparse CO coverage.

The shift in the temperature at which the rate of CO_2 production on palladium (111) reaches its maximum value (T_{max}) to higher values at higher CO pressures can be understood by reference to (1) and (2). In the high temperature region where θ_{CO} is small, the uptake increases with increase in pressure. A higher temperature is thus required to overcome the inhibiting effect of the adsorbed CO. The fact that there is an increase in the rate of CO_2 production at the maximum can also be understood qualitatively. At the higher temperatures the velocity constant for reaction on the surface is expected to increase in the normal way.

2. Pressure dependence of oxidation rate

When we come to consider the kinetics of the reactions with respect to the gas phase pressures we find that the results depend upon the tempera-

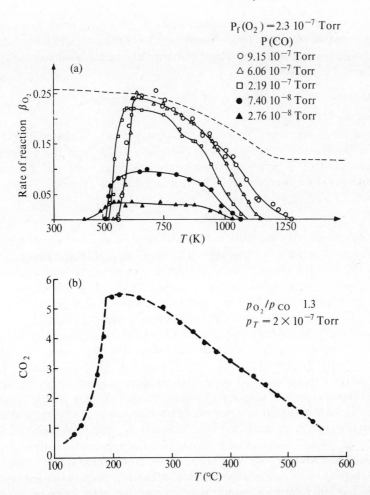

Fig. 9.23. (a) Steady state oxidation of CO on polycrystalline platinum as measured by the rate of consumption β_{O_2} of oxygen. s_O for oxygen on clean platinum is also shown (- - -). This is the same catalyst as that used for the results shown in Fig. 9.19. Above 1250 K s_O is due to the pumping effect of hot platinum and not to the adsorption of oxygen. (From Pacia et al. 1976.) (b) Steady state oxidation of CO on platinum (110) as a function of the platinum temperature. (From Bonzel and Ku 1972b.)

ture, the individual gas pressures, and the pressure ratio p_{CO}/p_{O_2}. No single general algebraic formulation of the results is practicable. Some of the limiting cases where reasonably clear-cut kinetic results are available will now be considered.

Palladium

1. $700 \, K > T > T_{max}$

Above the temperature T_{max} of the maximum rate on palladium (111) but below 700 K, the oxygen layer is thermally stable and the equilibrium uptake of CO is sparse. For catalytic oxidation by the Langmuir–Hinshelwood mechanism, the rate in this region is given by

$$R_{LH} = k_4 \theta_O \theta_{CO}. \tag{9.10}$$

The quantities with the most substantial dependence on temperature are k_4 and θ_{CO}, which are proportional to $\exp(-E_{LH}/RT)$ and $\exp(E_{des}/RT)$ (eqn (9.8)) respectively. The rate should thus decrease with increase in temperature, since as we have already seen $E_{des} > E_{LH}$. This is in accord with observation, although the apparent activation energy is not exactly $|E_{des} - E_{LH}|$ since θ_O shows a modest temperature dependence.

The kinetic behaviour on palladium (111) in this region depends upon whether the ratio p_{O_2}/p_{CO} exceeds unity or is less than unity. We can see broadly why this should be so, if we note first that the sticking probabilities of oxygen and CO on a sparsely covered surface are about 0.4 and 1 respectively. Thus, equal pressures of CO and oxygen produce roughly equal adsorption rates of CO molecules and O atoms, thereby generating the correct surface stoichiometry for the production of CO_2. However, whilst an excess of gaseous CO produces rather little change in the surface CO, since the equilibrium coverage is low at these temperatures, an excess of gaseous oxygen does affect θ_O. Surface oxygen is stable up to $T > 700 \, K$ and the superfluous oxygen can thus increase θ_O.

Let us now discuss the kinetic implications of these considerations.

(a) $p_{O_2} < p_{CO} < 10^{-6}$ Torr ($T > T_{max}$). In this situation the surface is sparsely covered with O(ad) and CO(ad) as illustrated in Fig. 9.10. The CO(ad) is in excess, although it is not sufficient to lead to catalytic inhibition. The rate of CO_2 formation is found to be essentially independent of the pressure of CO. This observation can be attributed to a surface reaction, (eqn (9.10)) so rapid that it is not the rate-determining step. Under these circumstances the rate of reaction is determined by the rate of adsorption of oxygen, i.e.

$$R = 2k_3 p_{O_2}. \tag{9.11}$$

The results for a temperature of 522 K and a CO pressure of 1×10^{-7}

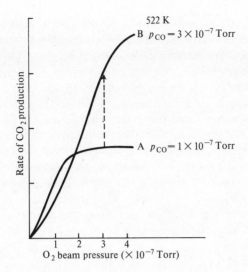

Fig. 9.24. Catalytic oxidation by palladium (111). Steady state production of CO_2 on palladium (111) at 522 K at two isotropic CO pressures as a function of the oxygen beam (equivalent) pressure: $-$ $-$, restoration of first-order kinetics with respect to the oxygen pressure (From Engel and Ertl 1978.)

Torr are shown in Fig. 9.24. As can be seen, provided that $p_{CO} > p_{O_2}$, the rate of CO_2 production does rise essentially linearly with p_{O_2} as required by eqn (9.11). However, when the pressure of oxygen increases to the point where $p_{CO} < p_{O_2}$, as in the upper reaches of the curve (A) of Fig. 9.24, oxygen adsorption is no longer rate determining. A limiting rate is then observed.

(b) $p_{O_2} > p_{CO} < 10^{-6}$ Torr $(T > T_{max}, \theta_O > 0.1)$. Under these conditions where the coverage of CO is low, the reaction on palladium (111) reaches the limiting rate, independent of p_{O_2}, discussed in the previous section. An increase in p_{CO} now increases the rate of CO_2 production and eventually restores the reaction to its first-order dependence on p_{O_2}, as shown for example in Fig. 9.24 by the transition from curve A to curve B at $p_{O_2} = 3 \times 10^{-7}$ Torr.

There is an interesting feature in the kinetics with respect to the surface coverage of oxygen; not only is the reaction independent of p_{O_2}, it is also independent of θ_O for $\theta_O > 0.1$. This result is not expected for the simple Langmuir–Hinshelwood mechanism. The interpretation comes from the structural studies discussed earlier. LEED data showed that in this coverage region the oxygen atoms gave the (2×2) structure. Domains with $\theta_O = \frac{1}{4}$ are required for this structure, so that for $\theta_O \approx 0.1$, there must also be bare patches on the surface. In effect, therefore, for $\theta_O \approx 0.1$ there are islands of oxygen atoms separated by areas of clean surface. CO molecules may adsorb on these latter areas. As we noticed previously, adsorbed CO is much more mobile on the surface than adsorbed oxygen is. To a first approximation therefore, the formation of CO_2 can be visualized as taking place by the collision of a freely moving chemisorbed CO molecule with an immobile O atom at the edge of an island. As long as the residence time of the CO molecules on the surface allows them to travel a distance which is long compared with the inter-island separation, the rate of CO_2 production is independent of oxygen coverage. However, if on increasing the temperature a point is reached at which some CO molecules desorb before reaching an oxygen island, the situation is changed. Now an increase in oxygen coverage increases the chance of CO colliding with an O atom, and the rate goes up. The kinetics then tend towards a linear dependence on θ_O as well as on θ_{CO}. In this region the rate once again follows the Langmuir–Hinshelwood equation:

$$R \propto k_4 \theta_O \theta_{CO}.$$

Substitution of $\theta_{CO} \propto p_{CO} \exp(E_{des}/RT)$ (eqn (9.7)) and $k_4 = \nu_4 \exp(-E_{LH}/RT)$ gives

$$R \propto \theta_O p_{CO} \nu_4 \exp\left(\frac{E_{des} - E_{LH}}{RT}\right).$$

The measurement of the temperature dependence of R at constant θ_O yields a value of E_{LH} equal to 105 kJ mol^{-1}, in satisfactory agreement with the earlier value obtained under different experimental conditions.

Having thus examined some aspects of the kinetic behaviour when the CO coverage is sparse let us now examine what happens at lower temperatures where θ_{CO} may be appreciable.

2. $T < T_{max}$

In this temperature range the overall temperature dependence of the rate is largely governed by the desorption of CO, and thus has an apparent activation energy of about E_{des}. As far as the Langmuir–Hinshelwood surface reaction is concerned, a simple temperature dependence of its rate is not to be expected. Co-adsorption of CO and oxygen at appreciable coverages compresses the adsorbed oxygen layer, reduces its binding energy, and thus reduces E_{LH}. The result is an activation energy for the surface reaction which is reduced below the value at $T > T_{max}$ by an amount which depends on coverage. We have already seen this effect in Fig. 9.18 in which the lower temperature section of the graph has a slope of $E_{LH} = 59$ kJ mol^{-1}, whilst at higher temperatures the low CO coverage value of E_{LH} was 113 kJ mol^{-1}.

Turning now to the kinetic aspects, we note that, as shown in Fig. 9.22 the rate of CO_2 production at any particular temperature below T_{max} drops sharply as p_{CO} is increased in the range 1×10^{-7}–1×10^{-6} Torr. This result is anticipated from the inhibition by surface CO of the adsorption of oxygen. An approximate rate law in this low temperature region which shows the inhibiting effect of CO is

$$R \approx k' \frac{p_{O_2}}{p_{CO}}.$$

As before, the qualitative reason for the shift of T_{max} to higher temperatures as p_{CO} increases can be understood. The additional inhibiting effect of CO, illustrated by p_{CO} appearing in the denominator of the rate equation, at increased pressures must be offset by a higher temperature.

Platinum

The steady state oxidation of CO on platinum has broadly similar features to those discussed for the reaction on palladium. We shall therefore deal with them in less detail.

We start with a consideration of the measurements shown in Fig. 9.23(a) for which oxygen formed a molecular beam of constant equivalent pressure 2.3×10^{-7} Torr whilst CO was at an isotropic pressure between 2.7×10^{-8} and 9×10^{-7} Torr.

The first point to note is that the higher was the pressure of CO the higher was the temperature at which the rate of the reaction became measurable. Secondly, once started the reaction rate rose steeply with increase in catalyst temperature. Finally, it can be seen that a further increase in temperature eventually reduced the rate to zero.

All of these observations are similar to the behaviour of palladium and are in accord with a Langmuir–Hinshelwood mechanism. They can therefore be interpreted as a consequence of (i) catalytic inhibition by CO at the lower temperatures, (ii) a rise of rate when CO desorption sets in, and (iii) a decline as $\theta_{CO} \to 0$ at the highest temperatures.

The Langmuir–Hinshelwood mechanism also gives an account of the rise in the rate (β_{O_2}) of consumption of oxygen at T_{max} as p_{CO} increased. Since T_{max} was high ($\sim$650 K) the equilibrium coverage of CO was low, but increased with p_{CO}. Accordingly, the rate of reaction increased. Eventually, however, the reaction rate became limited by the availability of oxygen, rather than of CO, so that the upper limit to β_{O_2} became the sticking probability of oxygen. This situation is realized at the peak of the highest p_{CO} curves shown in Fig. 9.23(a).

The reactivity of the single-crystal platinum (110) followed the same pattern as the polycrystalline sample. The temperature of the peak rate increased at higher CO pressures for precisely the same reason as the increase on palladium (111) discussed earlier. This surface also illustrated clearly the dependence of the kinetic behaviour on the ratio p_{O_2}/p_{CO}. With oxygen in excess ($p_{O_2}/p_{CO} > 1$) the rate of production of CO_2 increased as p_{CO} increased, but for CO in excess ($p_{O_2}/p_{CO} < 1$) the converse was true. The results for the catalyst at the peak temperature (212 °C) and $p_{O_2} = 5.6 \times 10^{-8}$ Torr are illustrated in Fig. 9.25. The kinetic data can be summarized as

$$p_{O_2} > p_{CO}: \quad R_{CO_2} \propto p_{CO}^{0.75}$$
$$p_{O_2} < p_{CO}: \quad R_{CO_2} \propto p_{CO}^{-0.6}$$

In experiments where p_{CO} was kept constant and p_{O_2} was varied it was found that when $p_{O_2} < p_{CO}$ the reaction rate was approximately first order in oxygen pressure, but that for $p_{O_2} > p_{CO}$ the rate was only slightly dependent on p_{O_2} ($p_{O_2}^{0.25}$).

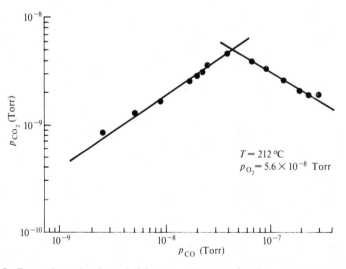

Fig. 9.25. Rate of production of CO_2 on a platinum (110) sample at 212 °C as a function of the CO pressure. The initial oxygen pressure is constant at 5.6×10^{-8} Torr. (From Pacia *et al.* 1976.)

Reactivity of supported palladium crystallites

There is satisfactory agreement between the results we have discussed in some detail for the palladium (111) plane and those obtained at low pressures for other palladium planes. A summary of the steady state rates of CO_2 formation is shown in Fig. 9.26. More striking perhaps is the concordance with the results of a study of catalysis by crystallites of palladium evaporated onto an alumina support. In this elegant experiment a thin single crystal of alumina prepared under ultrahigh vacuum was used as the support. The palladium was evaporated onto the substrate in a controlled way such that crystallites of a uniform diameter ranging from 15 to 80 Å depending on the conditions were produced. The diameters were measured by transmission electron microscopy of an ultrathin area of the support and thus without the need to remove the sample from the ultrahigh vacuum chamber. Bright field micrographs of large and small particles are shown in Fig. 9.27.

The temperature dependence of the rate of oxidation was independent of crystallite size and was similar to that observed on palladium (111) and illustrated in Fig. 9.22. The results on crystallites of diameter 4.9 nm are shown for two p_{O_2}/p_{CO} ratios at constant $p_{O_2} = 1 \times 10^{-7}$ Torr in Fig. 9.28.

The kinetic observations on the crystallites are summarized in Table 9.1.

The temperature dependence and kinetic results agree entirely with

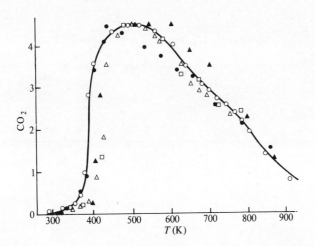

Fig. 9.26. The steady state rate of production of CO_2 (in arbitrary units) as a function of temperature for a variety of palladium surfaces: $\triangle$, palladium (111); $\bullet$, palladium (100); $\square$, palladium (110); $\blacktriangle$, palladium (210); $\bigcirc$, palladium wire. (From Ertl 1980.)

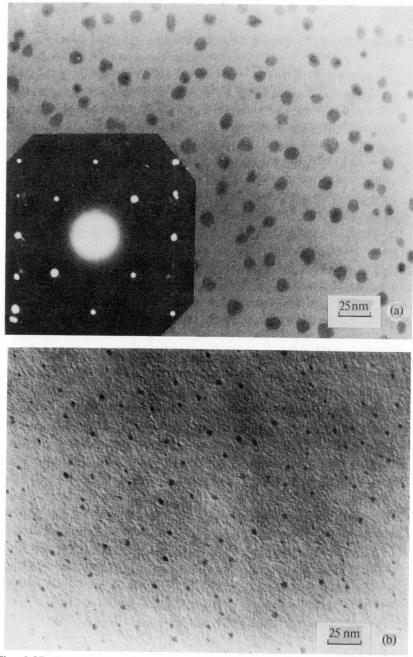

Fig. 9.27. Bright field micrographs of two transmission electron microscopy specimens of palladium crystallites on single-crystal α-Al$_2$O$_3$. Average crystallite diameters are (a) 8 nm and (b) 1.9 nm. The insert in (a) shows the transmission electron diffraction pattern. The strong spots are due to α-Al$_2$O$_3$, the faint ones are from palladium particles. (From Ladas, Poppa, and Boudart 1981.)

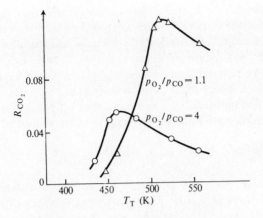

Fig. 9.28. Rate of production of CO_2 on palladium crystallites (4.9 nm in diameter) as a function of substrate temperature ($p_{O_2} = 1 \times 10^{-7}$ Torr). (From Ladas *et al.* 1981.)

Table 9.1

Temperature range	Pressure range	Kinetics
$T < T_{max}$	constant $p_{O_2} > p_{CO}$	$\dfrac{d[CO_2]}{dt} \propto \dfrac{1}{p_{CO}}$
	constant $p_{CO} > p_{O_2}$	$\dfrac{d[CO_2]}{dt} \propto p_{O_2}$
	constant p_{CO}/p_{O_2} (limited T range)	$\dfrac{d[CO_2]}{dt}$ independent of p_{total}
$T > T_{max}$	constant $p_{O_2} > p_{CO}$	$\dfrac{d[CO_2]}{dt} \propto p_{CO}$
	constant $p_{O_2} < p_{CO}$	$\dfrac{d[CO_2]}{dt}$ independent of p_{CO}
	constant $p_{CO} > p_{O_2}$	$\dfrac{d[CO_2]}{dt} \propto p_{O_2}$
	constant $p_{CO} < p_{O_2}$	$\dfrac{d[CO_2]}{dt}$ independent of p_{O_2}
	constant p_{CO}/p_{O_2}	$\dfrac{d[CO_2]}{dt} \propto p_{total}$

Kinetic observations for CO oxidation on palladium crystallites.

earlier single-crystal data and demonstrate clearly that the reaction is structure insensitive. Finally, it is interesting to note the comparison with the reactivity of palladium supported on finely divided alumina and used as a catalyst for oxidation at pressures near 1 atm. For temperatures below 450 K the reactivity as expressed by the 'turnover rate' was the same' (to within a factor of 2) despite an increase in pressure by at least seven orders of magnitude. The reason suggested for this striking observation was that, as noted above, the rate is independent of p_{total}.

Epilogue

In conclusion it seems indubitable that the catalytic oxidation of CO on palladium (111) and probably other palladium surfaces proceeds by the Langmuir–Hinshelwood mechanism under all the experimental conditions so far investigated. However, as far as platinum is concerned, the rate data are even more complex than for palladium. The safe deduction may be that, whilst platinum generally follows palladium in its behaviour, the Langmuir–Hinshelwood mechanism may not have a complete monopoly. More will be heard on this subject.

References

Atkins, P. W. (1982). *Physical chemistry*, 2nd Ed., Oxford University Press.
Barteau, M. A., Ko, E. I., and Madix, R. J. (1981). *Surf. Sci.*, **102,** 99.
Behm, R. J., Christmann, K., and Ertl, G. (1980). *Surf. Sci.* **99,** 320.
—— —— —— Van Hove, M. A. (1980). *J. Chem. Phys.* **73,** 2984.
—— —— —— —— Thiel, P. A. and Weinberg, W. H. (1979). *Surf. Sci.* **88,** L59.
Bell, R. P. (1980). *The tunnel effect in chemistry*. Chapman and Hall, London.
Bell, A. T. and Hair, M. L. (eds.) (1980). *Vibrational spectroscopies for adsorbed species. ACS Symp. Ser.* 137. American Chemical Society, Washington, DC. *ACS Symp. Ser.* 137. American Chemical Society, Washington, DC.
Berning, G. L. P., Allredge, G. P., and Viljoen, P. E. (1981). *Surf. Sci.* **104,** L225.
Bond, G. C. (1962). *Catalysis by metals*. Academic Press, New York.
Bonzel, H. P., Franken, A. M., and Pirug, G. (1981). *Surf. Sci.* **104,** 625.
—— and Ku, R. (1972a) *Surf. Sci.* **33,** 91.
—— —— (1972b) *J. Vac. Sci. Technol.* **9,** 663.
Boudart, M. (1976). *Proc. 6th Int. Congr. on Catalysis*, Vol. 1, p. 1. Chemical Society, London, 1977.
Bowker, M. and King, D. A. (1979). *J. Chem. Soc., Faraday Trans. I* **75,** 2100.
Bradshaw, A. M. (1979). *Surf. Sci.* **80,** 215.
Brucker, C. F. and Rhodin, T. N. (1979). *Surf. Sci.* **86,** 638.
Campbell, C. T., Ertl, G., Kuipers, H., and Segner, J. (1980). *J. Chem. Phys.* **73,** 5862.
—— —— —— —— (1981). *Surf. Sci.* **107,** 207.
Cassuto, A. and King, D. A. (1981). *Surf. Sci.* **102,** 388.
Conrad, H., Ertl, G., and Küppers, J. (1978). *Surf. Sci.* **76,** 323.
Cotton, F. A. and Wilkinson, G. (1980). *Advanced inorganic chemistry*, 4th edn., John Wiley, New York.
Dubois, L. H. and Somorjai, G. A. (1980). In *Vibrational spectroscopies for adsorbed species* (eds. A. T. Bell and M. L. Hair). *ACS Symp. Ser. 137.* American Chemical Society, Washington, DC.
Edwards, S. M., Gasser, R. P. H., Green, D. P., Hawkins, D. S., and Stevens, A. J. (1978). *Surf. Sci.* **72,** 213.
Ehrlich, G. (1980). *J. Vac. Sci. Technol.* **17,** 9.
Eland, J. H. D. (1974). Photoelectron spectroscopy. Butterworths, London.
Eley, D. D. and Norton, P. R. (1966). *Disc. Faraday Soc.* **41,** 135.
Engel, T. W. (1978). *J. Chem. Phys.* **69,** 373.
—— and Ertl, G. (1978). *J. Chem. Phys.* **69,** 1267.
—— —— (1979). *Adv. Catal.* **28,** 2.
Ertl, G. (1980). *Pure Appl. Chem.* **52,** 2051.
Feuerbacher, B., Fitton, B., and Willis, R. F. (1978). *Photoemission and the electronic properties of surfaces*. Wiley-Interscience, Chichester.
Ford, R. R. (1970). *Adv. Catal.* **21,** 51.
Gasser, R. P. H. and Holt, D. E. (1977). *Surf. Sci.* **64,** 520.
—— Morton, T. N., Overton, J. M., and Szczepura, A. K. (1971). *Surf. Sci.* **28,** 574.

—— and Perry, G. J. (1979). *J. Catal.* **60,** 378.

—— and Richards, W. G. (1974). *Entropy and energy levels.* Clarendon Press, Oxford.

Gates, B. C., Katzer, J. R. and Schuit, G. C. A. (1979). Chemistry of catalytic processes. McGraw Hill, New York.

Gimblett, F. G. R. (1970). *Introduction to the kinetics of chemical chain reactions.* McGraw Hill, London.

Gomer, R. (1961). *Field emission and field ionization. Harvard monographs in applied science 9.* Harvard University Press, Cambridge, MA.

—— Hulm, J. K. (1957). *J. Chem. Phys.* **27,** 1363.

Gorte, R. and Schmidt, L. D. (1978), *Surf. Sci.* **76,** 559.

Goymour, C. G. and King, D. A. (1973). *J. Chem. Soc., Faraday Trans. I* **69,** 736, 749.

Gregg, S. J. and Sing, K. S. W. (1967). *Adsorption, surface area and porosity.* Academic Press, London and New York. [Gives a thorough discussion of adsorption isotherms.]

Hayward, D. O., King, D. A. and Tompkins, F. C. (1967). *Proc. R. Soc. London, Ser. A,* **297,** 305.

Janda, K. C., Hurst, J. E., Becker, C. A., Cowin, J. P., Wharton, L., and Auerbach, D. J. (1980). *Surf. Sci.* **93,** 270.

King, D. A. and Wells, M. G. (1972). *Surf. Sci.* **29,** 454.

—— and —— (1974). *Proc. R. Soc. London, Ser. A* **339,** 245.

Kisliuk, P. (1957). *J. Phys. Chem. Solids* **3,** 95.

—— (1958). *J. Phys. Chem. Solids* **5,** 78.

Kramer, H. M. and Bauer, E. (1981). *Surf. Sci.* **107,** 1.

Kuijers, F. J. and Ponec, V. (1978). *Appl. Surf. Sci.* **2,** 431.

Küppers, J., Conrad, H., Ertl, G., and Latta, E. E. (1974). *Jpn. J. appl. Phys. Supp. 2,* Pt 2, 225.

Ladas, S., Poppa, H., and Boudart, M. (1981). *Surf. Sci.* **102,** 151.

Lambert, J. D. (1977). *Vibrational and rotational relaxation in gases.* Clarendon Press, Oxford, 1977. [Gives a lucid account of energy transfer in gases.]

Levy, R. A. *Principles of solid state physics* (1968). Academic Press, London.

Logan, R. M. and Stickney, R. E. (1966). *J. Chem. Phys.* **44,** 195.

—— and Keck, J. C. (1968). *J. Chem. Phys.* **49,** 860.

McCarroll, B. and Ehrlich, G. (1963). *J. Chem. Phys.* **38,** 523.

Madey, T. E. (1973). *Surf. Sci.* **36,** 281.

Nicholas, J. F. (1965). *An atlas of models of crystal surfaces.* Gordon and Breach, New York.

Niehus, H. and Comsa, G. (1981). *Surf. Sci.* **102,** L14.

Pacia, N., Cassuto, A., Pentenero, A., and Weber, B. (1976). *J. Catal.* **41,** 455.

Passler, M., Ignatiev, A., Jona, F. P., Jepsen, D. W., and Marcus, P. M. (1979). *Phys. Rev. Lett.* **43,** 360.

Pendry, J. B. (1974). *Low energy electron diffraction.* Academic Press, New York.

Pilling, M. J. (1975). *Reaction kinetics.* Clarendon Press, Oxford.

Pritchard, J. S. and Sims, M. L. (1970). *Trans. Faraday Soc.* **66,** 427.

Redhead, P. A. (1961). *Trans. Faraday Soc.* **57,** 641.

Rhodin, T. N. and Ertl, G. (1979). *The nature of the surface chemical bond.* North-Holland, Amsterdam.

Roberts, M. W. (1977). *Chem. Soc. Rev.* **6,** 373.

—— and McKee, C. S. (1978). *Chemistry of the metal–gas interface,* Clarendon Press, Oxford.

Rootsaert, W. J. M. and Sachtler, W. M. H. (1960). *Z. Phys. Chem. N. F.* **26,** 16.

Satterfield, C. N. (1980). *Heterogeneous catalysis in practice.* McGraw Hill, New York.

Shigeishi, R. A. and King, D. A. (1976). *Surf. Sci.* **58,** 379.

Solymosi, F. and Kiss, J. (1981). *Surf. Sci.* **108,** 368.

Somorjai, G. A. (1977). *Adv. Catal.* **26,** 2.

—— (1979). *Surf. Sci.* **89,** 496.

Stevens, M. A. and Russell, G. J. (1981). *Surf. Sci.* **104,** 354.

Strozier, J. A. (1979). *Surf. Sci.* **87,** 161.

Tamm, P. W. and Schmidt, L. D. (1970). *J. Chem. Phys.* **52,** 1150.

Tompkins, F. C. (1978). *Chemisorption of gases on metals.* Academic Press, London.

Turner, D. W. and May, D. P. (1966). *J. Chem. Phys.* **45,** 471.

Van Hove, M. A. and Tong, S. Y. (1979). *Surface crystallography by LEED.* Springer, Berlin.

Weinberg, W. H. and Merrill, R. P. (1971). *J. Vac. Sci. Technol.* **8,** 718.

Williams, F. L. and Nason, D. (1974). *Surf. Sci.,* **45,** 377.

Willis, R. F., Ho, W., and Plummer, E. W. (1979). *Surf. Sci.* **80,** 593.

Yates, J. T., Jr., Duncan, T. M., Worley, S. D., and Vaughan, R. W. (1979). *J. Chem. Phys.* **70,** 1219.

Index

Note: specific references to illustrations and tables are indicated by *italic numbers*.